D1031653

Sharing Transboundary Resources

International Law and Optimal Resource Use

Why do states often fail to cooperate, using transboundary natural resources inefficiently and unsustainably? Eyal Benvenisti examines the contemporary international norms and policy recommendations that could provide incentives for states to cooperate. His approach is multidisciplinary, proposing transnational institutions for the management of transboundary resources. Benvenisti takes a fresh approach to the problem, considering mismanagement as the link between domestic and international processes. As well, he explores reasons why some collective efforts to develop the international law on transnational ecosystems have failed, while others have succeeded. This inquiry suggests that adjudicators need to be assertive in progressively developing the law, while relying on scientific knowledge more than on past practice. Global water policy issues seem set to remain a cause for concern for the foreseeable future; this study provides a new approach to the problem of fresh water, and will interest international environmentalists and lawyers, as well as international relations scholars and practitioners.

EYAL BENVENISTI is Hersch Lauterpacht Professor of International Law and Director of the Minerva Center for Human Rights at the Hebrew University of Jerusalem. His publications include *The International Law of Occupation* (1993), as well as articles in the *European Journal of International Law*, the *American Journal of International Law*, and the *Michigan Law Review*.

CAMBRIDGE STUDIES IN INTERNATIONAL AND COMPARATIVE LAW

This series (established in 1946 by Professors Gutteridge, Hersch Lauterpacht and McNair) is a forum of studies of high quality in the fields of public and private international law and comparative law. Although these are distinct legal subdisciplines, developments since 1946 confirm their interrelationship.

Comparative law is increasingly used as a tool in the making of law at national, regional, and international levels. Private international law is increasingly affected by international conventions, and the issues faced by classical conflicts rules are increasingly dealt with by substantive harmonization of law under international auspices. Mixed international arbitrations, especially those involving state economic activity, raise mixed questions of public and private international law. In many fields (such as the protection of human rights and democratic standards, investment guarantees and international criminal law) international and national systems interact. National constitutional arrangements relating to "foreign affairs," and to the implementation of international norms, are a focus of attention.

Professor Sir Robert Jennings edited the series from 1981. Following his retirement as General Editor, an editorial board has been created and Cambridge University Press has recommitted itself to the series, affirming its broad scope.

The Board welcomes works of a theoretical or interdisciplinary character, and those focusing on new approaches to international or comparative law or conflicts of law. Studies of particular institutions or problems are equally welcome, as are translations of the best work published in other languages.

General Editors	Professor James R. Crawford SC FBA *Whewell Professor of International Law, Faculty of* *Law and Director, Lauterpacht Research Centre for* *International Law University of Cambridge* Professor John S. Bell FBA *Professor of Law* *Faculty of Law* *University of Cambridge*
Editorial Board	Professor Hilary Charlesworth, *University of Adelaide* Professor Lori Damrosch, *Columbia University Law School* Professor John Dugard, *Universiteit Leiden* Professor Mary-Ann Glendon, *Harvard Law School* Professor Christopher Greenwood, *London School of Economics* Professor David Johnston, *University of Edinburgh* Professor Donald McRae, *University of Ottawa* Professor Onuma Yasuaki, *University of Tokyo* Professor Reinhard Zimmermann, *Universität Regensburg*
Advisory Committee	Professor Sir D. W. Bowett QC Judge Rosalyn Higgins QC Professor Sir Robert Jennings QC Professor J. A. Jolowicz QC Professor Sir Eli Lauterpacht QC Professor Kurt Lipstein QC Judge Stephen Schwebel

A list of the books in the series can be found at the end of this volume

Sharing Transboundary Resources

International Law and Optimal Resource Use

Eyal Benvenisti

Hebrew University of Jerusalem

PUBLISHED BY THE PRESS SYNDICATE OF THE UNIVERSITY OF CAMBRIDGE
The Pitt Building, Trumpington Street, Cambridge, United Kingdom

CAMBRIDGE UNIVERSITY PRESS
The Edinburgh Building, Cambridge CB2 2RU, UK
40 West 20th Street, New York, NY 10011-4211, USA
477 Williamstown Road, Port Meilbourne, VIC 3207, Australia
Ruiz de Alarcón 13, 28014 Madrid, Spain
Dock House, The Waterfront, Cape Town 8001, South Africa

http://www.cambridge.org

© Eyal Benvenisti 2002

First published 2002

Printed in the United Kingdom at the University Press, Cambridge

Typeface Swift 10/13 pt.   *System* LATEX 2_ε [TB]

A catalogue record for this book is available from the British Library

ISBN 0 521 64098 9 hardback

For Rivka

Contents

Acknowledgments

When I started this project almost ten years ago my goal seemed straightforward: I wanted to examine what guidelines international law offered for the resolution of the conflict over the use of the shared freshwater resources in the Middle East. Increasing scarcity on the one hand, and negotiations towards a peaceful settlement of the Arab–Israeli conflict on the other, lent a sense of urgency even to such an academic exercise. But soon I discovered that the basic principles of international law on shared fresh water were disputed, also by the governments in the Middle East, and that therefore negotiators were not significantly constrained by the law. Trying to explain why the law failed to offer shared norms I expanded my initial inquiry and focused on two related questions. The first dealt with substance: just what *ought to be* the law to govern shared fresh water. The second question was institutional: why the protracted efforts to codify the law failed to reduce its ambiguities. In responding to these questions I ventured into new disciplines. The study of hydrology taught me that I should focus on ecosystems rather than on watercourses. The study of game theory and public choice theory permitted me to explore the roots of the failure of collective action in the national, regional, and international arenas. This book is the outcome of this inquiry.

The book is therefore not only about the international law on transboundary ecosystems. It is also about the logic of regional cooperation in the management of such ecosystems, and the potential contribution of international law towards such cooperation. The book also offers a critical analysis of the processes that shape the evolution of international law, and especially on its conscious development by international adjudicators. This theoretical study ultimately yields practical recommendations on how to structure the mechanisms for the management of transboundary resources.

Some of my initial thoughts on these questions were published as an article "Collective Action in the Utilization of Shared Freshwater: The Challenges of International Water Resources Law," in the *American Journal of International Law* 384 (1996). An article "Exit and Voice in the Age of Globalization," that appeared in *Michigan Law Review* 167 (1999), and a book chapter in *The Role of Law in International Politics*, edited by Michael Byers (2000) focused on the interface between international law and domestic politics, an issue that is elaborated here in chapters 3 and 4.

My work benefited from the comments, criticisms, and suggestions of friends and colleagues. They include José E. Alvarez, Andrea Bianchi, Kathryn Comerford, Tsili Dagan, Michael Dorf, George Downs, Robert Ellickson, Celia Fassberg, Ruth Gavison, Zohar Goshen, Alon Harel, Moshe Hirsch, Marcel Kahan, Petros Mavroidis, Georg Nolte, Ariel Porat, Mark Ramseyer, Michael Reisman, Edward Rock, Charles Sabel, Alex Stein, Eyal Zamir, and Andreas Zimmermann. Haim Gvirtzman introduced me to the basics of hydrology. From Eran Feitelson I learnt about conservation and management of scarce resources.

I am indebted to a number of academic institutions for their collegial environments and outstanding libraries that proved crucial for my work. I spent a semester in 1996 as a Visiting Fellow in the Max Planck Institute for Comparative Public Law and International Law in Heidelberg, a year (1998/9) as Visiting Professor at Harvard Law School, and three semesters as a Visiting Professor at Columbia Law School (1999, 2000, 2001). My friends at the Law Faculty of The Hebrew University of Jerusalem offered a constantly stimulating intellectual community.

Support for my project was generously provided by the Israel Science Foundation, the United States Institute of Peace, and the Bruno Goldberg Foundation. The Fulbright–Yitzhak Rabin Award, granted by the United States–Israel Educational Foundation in 1998, enabled me to devote precious time for research and writing. I am most grateful to these institutions.

Lastly, I wish to express my gratitude to my wife Rivka and our sons, Haggai and Amir, for their patience, encouragement, and inspiration.

Table of cases

International Court of Justice

(including the Permanent Court of International Justice)

Arbitral awards

WTO Appellate Body reports

European Court of Justice

Selected United States Cases

Selected cases of other national courts

Abbreviations

AJIL – *American Journal of International Law*
Am. Ar. Aff. – *American Arab Affairs*
Am. Jur. Pol. Sci. – *American Journal Of Political Science*
Am. Pol. Sci. Rev. – *American Political Science Review*
Ann. Franc. Dr. Int'l – *Annuaire français de droit international*
Ann. Inst. Dr. Int'l – *Annuaire de l'institut de droit international*
Ariz. J. Int'l & Comp. Law – *Arizona Journal of International and Comparative Law*
Austrian J. Publ. Int'l Law – *Austrian Journal of Public International Law*
ARIEL – *Austrian Review of International and European Law*
Bell J. Econ & Mngm't Sci. – *Bell Journal of Economics and Management Science*
Brit. J. Pol. Sci. – *British Journal of Political Science*
Brit. Yb. Int'l L. – *British Yearbook of International Law*
Brook. J. Int'l L. – *Brooklyn Journal of International Law*
Calif. L. Rev. – *California Law Review*
Cal. W. Int'l L.J. – *California Western International Law Journal*
Cambridge J. of Econ. – *Cambridge Journal of Economics*
Can. Yb. Int'l L. – *Canadian Yearbook of International Law*
Cardozo L. Rev. – *Cardozo Law Review*
Chi. Kent L. Rev. – *Chicago–Kent Law Review*
Colo. J. Int'l Envt'l L. & Pol'y – *Colorado Journal of International Environmental Law and Policy*
Colum. J. Envt'l L. – *Columbia Journal of Environmental Law*
Colum. J. Trans. L. – *Columbia Journal of Transnational Law*
Colum. L. Rev. – *Columbia Law Review*
Cornell Int'l L.J. – *Cornell Journal of International Law*
Cornell L. Rev. – *Cornell Law Review*

Duke Env. L & Pol'y F. – *Duke Environmental Law and Policy Forum*
Duke L.J. – *Duke Law Journal*
Ecology L.Q. – *Ecology Law Quarterly*
Emory Int'l L. Rev – *Emory International Law Review*
Eur. J. Int'l L. – *European Journal of International Law*
Fordham L. Rev. – *Fordham Law Review*
Georgetown Int'l Envtl L. Rev. – *Georgetown International Environmental Law Review*
Golden Gate U.L. Rev. – *Golden Gate University Law Review*
Hague YB Int'l L. – *Hague Yearbook of International Law*
Harv. Envtl. L. Rev. – *Harvard Environmental Law Review*
Harv. Int'l L.J. – *Harvard Journal of International Law*
Harv. L. Rev. – *Harvard Law Review*
Hous. J. Int'l L. – *Houston Journal of International Law*
ICLQ – *International and Comparative Law Quarterly*
ILM – *International Legal Materials*
Ind. L.J. – *Indiana Law Journal*
Int'l J. Group. Rts – *International Journal on Group Rights*
Int'l L.R. – *International Law Reports*
Int'l Org. – *International Organization*
Int'l Sec. – *International Security*
Israel L. Rev. – *Israel Law Review*
J. Conflict Res. – *Journal of Conflict Resolution*
J. Econ. Theo. – *Journal of Economic Theory*
J. Int'l Aff. – *Journal of International Affairs*
J. Int'l Econ. L. – *Journal of International Economic Law*
J. Int'l L. Bus. – *Journal of International Law and Business*
J. Law & Econ. – *Journal of Law and Economics*
J. Law, Econ. & Org. – *Journal of Law, Economics and Organization*
J. Leg. Stud. – *Journal of Legal Studies*
J. Palestine Studies – *Journal of Palestine Studies*
J. Pol'y Analysis & Mgmt. – *Journal of Policy Analysis and Management*
J. Pol. Econ. – *Journal of Political Economy*
Mich. J. Int'l L. – *Michigan Journal of International Law*
Mich. L. Rev. – *Michigan Law Review*
Minn. J. Global Trade – *Minnesota Journal of Global Trade*
Minn. L. Rev. – *Minnesota Law Review*
NAFTA L. & Bus. Rev. Am. – *NAFTA: Law and Business Review of the Americas*
Nat. Res. J. – *Natural Resources Journal*

Netherlands Yb. Int'l L. – *Netherlands Yearbook of International Law*
Nordic J. Int'l L – *Nordic Journal of International Law*
Nw. U. L. Rev. – *Northwestern University Law Review*
NYU J. Int'l L. & Pol., – *New York University Journal of International Law and Policy*
NYU L. Rev. – *New York University Law Review*
Pace Envtl. L. Rev. – *Pace Environmental Law Review*
Q. J. Econ. – *Quarterly Journal of Economics*
Rev. Belge Dr. Int'l – *Revue belge de droit international*
Rev. Gen. Dr. Int'l Pub. – *Revue général de droit international public*
Rev. Jur. Envn't – *Revue Juridique de l'environnement*
S. Cal. L. Rev. – *Southern California Law Review*
Schweiz. Z. Int'l Europ. R. – *Schweizerische Zeitschrift fuer Internationales und Europaeisches Recht*
Stan. Envtl. L.J. – *Stanford Environmental Law Journal*
Stan. J. Int'l L. – *Stanford Journal of International Law*
Stan. L. Rev. – *Stanford Law Review*
Tex. Int'l L.J. – *Texas Journal of International Law*
Trans. Inst. Br. Geogr. – *Transactions of the Institute of British Geographers*
U. Chi. L. Rev. – *University of Chicago Law Review*
U. Cin. L. Rev. – *University of Cincinnati Law Review*
UCLA L. Rev. – *University of Southern California Law Review*
U. Col. L. Rev. – *University of Colorado Law Review*
U. Denv. Water L. Rev. – *University of Denver Water Law Review*
U Pa. J. Int'l Econ. L. – *University of Pennsylvania Journal of International Economic Law*
U. Pa. L. Rev. – *University of Pennsylvania Law Review*
U. Rich. L. Rev. – *University of Richmond Law Review*
U. West. Ontario L. Rev. – *University of Western Ontario Law Review*
Va. J. Int'l L. – *Virginia Journal of International Law*
Va. L. Rev. – *Virginia Law Review*
Vand. L. Rev. – *Vanderbilt Law Review*
Wash. & Lee L. Rev. – *Washington and Lee Law Review*
Yale HR & Dev. L.J. – *Yale Human Rights and Development Law Journal*
Yale J. Int'l L. – *Yale Journal of International Law*
Yale L.J. – *Yale Law Journal*
Yb. Int'l Envt'l L. – *Yearbook of International Environmental Law*

1 Introduction

Transboundary resources: delineating the challenges

For the thousands of Muslim worshippers who gathered in mosques across the Middle East one Friday morning, as the second millennium was drawing to an end, only God could end the misery caused by the worst drought experienced in their lifetimes. Thousands of Jewish worshippers joined them the following morning, fervently reciting the daily prayer for rain. Indeed, as these prayers suggested, the occurrence of drought was a matter beyond human control. Yet the praying, which the political leaders ceremoniously attended, furthered the wrong perception of water shortage as a problem of dwindling supplies. It de-emphasized the governments' responsibility for the inability to manage responsibly the conflicting demands for water and to reduce waste. Indeed, much of the plight of the worshippers was a result of human conflict and government failure to correct inefficiencies in water management systems and to prevent environmental degradation.

Dating back three millennia, the Middle East has been a region where impressive instances of efficient small-scale demand-management systems have thrived. Villagers have managed to design and implement collective mechanisms for the shared management of small springs, aquifers, and floods. Thanks to these ancient systems, many of these villages survive to this very day. One would have hoped that the emergence of the modern state in the Middle East towards the end of the second millennium would have produced similar successful arrangements on a regional or even national scale. But the governments in the Middle East have failed to do so and, instead, have caused much dissipation and ruin of natural resources. The picture is similar in other parts of the world: efficient small-scale water management institutions have been replaced by larger, inefficient, and often corrupt systems with consequential loss and even human suffering.

The same Middle Eastern leaders who joined the dramatic prayer for rain have also failed to ameliorate the dismal situation through negotiations with one another on resolving the regional water and environmental disputes. A comprehensive plan that could have reduced waste and increased water availability eluded them. Since gaining independence, the states in the Middle East have been engaged in conflict with one another, often with saber-rattling that led, on one occasion, in 1967, to all-out war. Needless to say, these conflicts have contributed to the plight the worshippers now plead to God to end.

When I first set out to explore the roots of the inefficiencies of the modern state and the causes of regional conflict in the management of transboundary natural resources, I was struck by the efficiency and sustainability of the ancient local systems. Why, I wondered, did villagers in ancient communities succeed where modern states fail? Are there systemic failures in the Westphalian state system that hinder efficient management of domestic and international resources? Can these systemic failures be corrected through institutions and norms? This book is the outcome of that endeavor to understand this perplexity. The most immediate goal of the book is to address the challenge of management of transboundary resources, namely natural resources shared by more than one state, in an efficient, sustainable, and equitable way. The book explores the reasons for inefficiency and non-sustainability, examines different responses that have been suggested, and proposes norms and institutions that could create more effective incentives for states to cooperate. On a more general level, this book provides a new outlook on the state as a locus of political decision making in the emerging global environment. It suggests that the principle of state sovereignty that allocates power to governments empowers some domestic interest groups at the expense of others. The difference in the aptitudes of the domestic groups to influence the ways the state manages its public resources often leads to inefficient and inequitable outcomes. The book argues that domestic and international norms and institutions can and should rectify this imbalance.

The focus of this book is on management of transboundary resources (also called "international common pool resources"). These are transboundary natural resources to which only a number of states have access. Such resources could be fresh water, clean air, fisheries in shared rivers and lakes, hydrocarbon and mineral deposits, forests and rainforests, nature reserves, and endangered species of flora and fauna. What characterizes these resources is their partial accessibility. Only a limited number of states enjoy access to the given resource. While their own access is unlimited, states can limit the access of other states. This opportunity to limit the access of others gives rise to the theoretical possibility that efficient and equitable collective action among the co-owners of the transboundary resource can emerge. I explore this theoretical possibility and suggest legal and institutional principles that could assist in accomplishing that possibility.

The rest of this introductory chapter is devoted to a more thorough clarification of the book's agenda and goals. It proceeds from the village level to the international level, identifying the main promises and pitfalls of collective use of common pool resources.

The endogenous evolution of cooperation in small-scale common pool resources

The collective effort to ensure an adequate supply of water was the bond that gave birth to many societies. Communities in arid and semi-arid areas had to coordinate activities to procure sufficient water to feed their families and cattle and to irrigate their fields. In other areas, where water was abundant, cooperation was necessary to prevent flooding. This endogenous cooperation resulted in efficient utilization of the communal resources. The design of sophisticated engineering projects could not have been sustained without equally sophisticated social, political, and legal designs. No well would be dug unless its water could be protected under a clearly defined set of rules of either individual or collective ownership. When the procurement of water required efforts beyond the capabilities of a single peasant, systems of common decision-making and monitoring were set up to collectively procure and apportion the shared resource.

The first story of successful cooperation is reported in the biblical tale of the meeting between Jacob and Rachel. A heavy stone covered the collective well that served the herds of all the villagers. Removing the stone required the joint effort of all the shepherds, but Jacob, in a show of extraordinary strength, managed to remove the stone single-handedly while trying to impress Rachel.[1] The heavy stone was a simple device that enabled collective monitoring of the timing and quantity of use, as

---

[1] Genesis 29:1–11.

well as assignment of responsibility for accidental pollution. The Middle East is replete with many similar examples, all based on the idea of a community-owned resource. One such example, which still functions, was developed in the ninth or eighth century BC. It involves a communal spring or system of springs. The villagers dug tunnels deep into the rock to drain the saturated aquifer more efficiently and increase the flow of these springs.[2] They based the complicated digging and maintenance of the spring flow tunnels and the distribution of the water thus obtained on the idea of the spring as a shared resource. A similar arrangement, which also emerged without the backing of a central government, developed through local customs in the ancient Persian kingdoms. Since the eighth century BC, farmers have irrigated their fields by groundwater flowing from *qanawat* (tunnels dug into the underground water table below riverbeds), which sometimes reached a length of more than fifty kilometers.[3] There is ample evidence that *qanawat* were satisfactorily operated, sometimes supplying over one hundred users.

Collective action required investment not only in infrastructure, but also in collective decision-making processes and enforcement mechanisms. In some communities, these functions depended heavily on family ties. The villagers in the Judean Hills in Palestine, for example, relied heavily on the structure of the *hammulah*, the extended family. Only a small number of *hammulahs* resided in each village, and water would rotate between the *hammulahs* on a weekly basis. At night, the spring water filled a publicly owned pool. Then, during the daytime, the water that had accumulated in the pool would be redirected to the fields, each day supplying water to the members of one *hammulah*. An elder of the *hammulah* would be in charge of the actual diversion. Zvi Ron described in detail the water system in Battir, an Arab village in the West Bank in the vicinity of Jerusalem, which, in 1967, still relied on the ancient spring flow allocation system.[4] Eight *hammulahs* lived in Battir,

---

[2] On the spring flow tunnels, see Zvi Y. D. Ron, "Qantas and Spring Flow Tunnels in the Holy Land" in Peter Beaumont, Michael Bonnie, and Keith McLachlan (eds.), *Qantas, Kariz and Khattara: Traditional Water Systems in the Middle East and North Africa* (London, Middle East & North African Studies Press, 1989), pp. 211–36. In some places, the tunnels reached a length of 50 to 100 meters and, in one place, even 225 meters (see at p. 224). See also "The Utilization of Springs for Irrigated Agriculture in the Judea Mountains," in Avshalom Shmueli, David Grossman and Rehav'am Ze'evi (eds.), *Judea and Samaria* (2 vols., Jerusalem, Canaan Publishing House, 1977, in Hebrew), vol. I, pp. 230–50.

[3] A. K. S. Lambton, "Qanat," 4 *Encyclopedia of Islam*, 529–31; Peter Beaumont, "The Qanat: A Means of Water Provision from Groundwater Sources" in Beaumont, Bonnie, and McLachlan, *Quantas*, note 2, pp. 13–31, at p. 23.

[4] Zvi Y. D. Ron, "Development and Management of Irrigation Systems in Mountain Regions of the Holy Land" (1985) 10 Trans. Inst. Br. Geogr. NS 149–69; Zvi Ron,

and hence, each *hammulah* would get water for its families every eighth day. An elder of the *hammulah* was in charge of distribution among the families of the *hammulah* and among the family members within each family. With a wooden stick that was notched with as many notches as there were water recipients, he would measure the decreasing water level in the pool and order the opening and closing of the pool gates. Throughout the day, several women from the same *hammulah* would sit near the pool, talking casually, but also watching the elder at work. Similar arrangements for collective processes of allocation of quantities and for monitoring actual withdrawals enabled indigenous populations in North America and in the Philippines to adjust to the sometimes harsh environment.[5]

Distribution in cycles provided a built-in response to fluctuations in water supply; when the source dwindled, everyone received less. Thus, maintenance of the spring and the nearby storage pool, as well as of the horizontal extension of the spring flow tunnels into the rock to capture more water, was in everyone's interest. This shared interest, backed by the reliable allocation system and enforced by the myriad of ties between and within families unrelated to water use,[6] enabled the development of long tunnels that extended well into the rock, well below the surface.

While strong family ties are conducive to reducing the costs of monitoring and enforcement, peasants in other regions have demonstrated that collective action can emerge despite the lack of such ties. Indeed, as Robert Wade has documented, fruitful cooperation emerged in some water-scarce villages in southern India, despite strict caste differences between the villagers and looser social ties.[7] Such cooperation developed

---

"Battir – The Village and the System of Irrigated Terraces" (1968) 10 *Teva va-Arets* 112, 121 (in Hebrew).

[5] For the irrigation systems of the Pueblo Indians in New Mexico, see Jose A. Rivera, "Irrigation Communities of the Upper Rio Grande Bioregion: Sustainable Use in the Global Context" (1996) 36 Nat. Res. J. 491, 497 (describing the "acequia associations," consisting of three elected ditch commissioners and the irrigators themselves, governed by rules based on custom and tradition); Robert Y. Siy, Jr., *Community Resource Management: Lessons from the Zanjera* (Quezon City, Philippines, University of the Philippines Press, 1982) (describing the irrigation system in rural parts of the Philippines).

[6] On multidimensional relations as reinforcing cooperation, see Elinor Ostrom, *Governing the Commons: The Evolution of Institutions for Collective Action* (New York, Cambridge University Press, 1990), p. 207; Russell Hardin, *Collective Action* (Baltimore, MD, Johns Hopkins University Press, 1982), pp. 31–3.

[7] Robert Wade, *Village Republics: Economic Conditions for Collective Action in South India* (Cambridge, Cambridge University Press, 1988); Robert Wade, "The Management of Common Property Resources: Collective Action as an Alternative to Privatisation or State Regulation" (1987) 11 Cambridge J. of Econ. 95.

due to the relative scarcity of the resource and the fact that the peasants held a diversified portfolio of fields: some near the water source, some further below, in the flatter area. The villagers described by Wade managed to form a council that coordinated the efforts to obtain more water for the village, monitored the allocation of this water, collected taxes to finance its actions, and fined violators. Violations occurred, and there were even suspicions that some farmers were using their position on the council to obtain unfair special benefits for themselves or their relatives. But all these concerns were addressed in public, on the local accountant's open veranda. Even more than fines did, the cost to reputation provided a reasonably effective sanction against violations. The council remained in operation for as long as it could ensure net gains to farmers from collective action.

But there were significant limits to these indigenous forms of cooperation. Both in the Judean Hills and in the southern Indian uplands, significant losses were caused by inter-village failure to cooperate. Sometimes the reason was the asymmetric upstream-downstream relationship between villages. At other times, due to sporadic seasonal flows, there was no incentive for setting up and maintaining mechanisms in anticipation for their occurrence. Here again, the Bible is a source of early evidence of conflict resulting from competition over water. The first biblical stories of conflicts in Canaan relate not to contested land, but to competition over access to water.[8] Lack of coordination often resulted not only in conflict, but also in inefficient and unsustainable use.[9]

Nevertheless, some communities managed to overcome even this collective failure. At times, religion proved a potent tool to iron out inter-village competition. Clifford Geertz describes this phenomenon, which survives to this day in parts of Indonesia.[10] In the Island of Bali, each

---

[8] Genesis 26:15–22 (the Philistines covered the wells dug by Abraham and Isaac in an attempt to chase Isaac away from the area).

[9] For examples of inter-village strife in ancient Palestine, see Ali Hasan Dawod Anbar, "Socio-Economic Aspects of the East Ghor Canal Project" (Ph.D. thesis, University of Southampton, 1983), pp. 91–3; for examples of inter-village conflicts in India, see Robert Wade, *Village Republics*, note 7. This is the typical scenario of the tragedy of the water commons and arises, for example, in areas along the Mediterranean coast, where the opportunity of many individuals to dig wells led to numerous shallow wells and a lowering of the water table which, in turn, rendered many wells dry and increased the salinity of the coastal aquifer.

[10] Clifford Geertz, "Organization of the Balinese Subak" in E. Walter Coward, Jr. (ed.), *Irrigation and Agricultural Development in Asia: Perspectives from the Social Sciences* (Ithaca, NY, Cornell University Press, 1980), pp. 70–90.

drainage basin has its own *subak*, or irrigation society. The *subak* is "in fact very much more: an agricultural planning unit, an autonomous legal corporation, and a religious community. Aside from house gardening, virtually everything having to do with cultivation lies within its purview."[11] It operates under a system that today could be described as subsidiarity, relegating decision making and activities as much as possible to the village level.[12] The bond between potentially rival villages has been the shared religion. "The begetter of order in this otherwise rather particulate social field is the temple system . . . The temple system provides both a simplified model of Balinese social structure and a schoolroom in which kinds of attitudes and values necessary to sustain it are inculcated and celebrated."[13] One of the three great temples, the Great Council Temple, holds an annual ceremony, which is the climax of lengthy preparations of representatives of the surrounding *subaks*. As Geertz observed, "the integrative force of this continual collective effort, as it moves from one social context to another, is the linchpin of the entire system."[14] This common belief system sustains an explicit local customary law that is enforced through negotiations.[15]

Often the policies of the emerging regimes in the developing world, supported by Western scientists irreverent to "native" and "primitive" cultures and practices, shattered those ancient systems. The modern systems, however, have proved less efficient. Contemporary scientists and disillusioned governments have now discovered that this and similar religious rites in Benin, Bolivia, and Cambodia may be more efficient than modern command and control systems run by short-sighted central bureaucracy, and strive to reconstruct them wherever this is still possible.[16] Since attempts to introduce modern strains of rice and fertilizers brought only environmental disaster, the Indonesian government has recently been trying to convince farmers to revert to the ancient Balinese "rice cult" noted so precisely by Geertz.[17]

---

[11] Geertz, "Organization of the Balinese Subak," p. 79.

[12] *Ibid.*: "Theories of 'hydraulic despotism' to the contrary notwithstanding, water control in Bali was an overwhelmingly local and intensely democratic matter." The *subak* encompasses all owners of rice fields irrigated by a single dam. Organization is based on a one-person one-vote system for electing the *subak* head and other officials who perform allocation, monitoring, and maintenance works (at pp. 80–1).

[13] *Ibid.*, at p. 81.   [14] *Ibid.*, at p. 88.   [15] *Ibid.*, at p. 81.

[16] Jane Ellen Stevens, "Science and Religion; Cultural Practices and Ecology" (1994) 44(2) *Bioscience* 60.

[17] *Ibid.*

The inefficiency and inequity of national command and control institutions

The temptation to ignore the limits of the local common pool resource and adopt an "economy-of-scale" approach is not a twentieth-century invention. Along with the successful small-scale efforts, the allure of water management on a grand scale was evident already in the ancient empires of Sumer and Assyria. Outsiders to the common resources collaborated with those who hoped to increase revenues from them by placing the smaller resources under an all-encompassing joint management. This required the replacement of the delicate mechanisms that had ensured individual incentives to cooperate with imposed rules and sanctions to compel cooperation. Such efforts gave birth to despotism and produced inefficient and unsustainable regimes. Disrespect for nature has been a major cause for the demise of such despotic empires.

This is the legacy of the ancient empires of Sumer and Assyria in Mesopotamia, described by Karl Wittfogel.[18] According to his account, in all great ancient civilizations, such as Sumer and Assyria in Mesopotamia, Pharaonic Egypt, the Inca Empire in Peru, ancient China, and India, taming the large rivers was the catalyst for their evolution. Wittfogel linked what he termed the "Oriental despotism" of these societies to their internal efforts to promote their economic growth through the use of more water to irrigate more fields. He therefore called them "hydraulic societies." Harnessing the mighty rivers for large-scale irrigation in fertile but otherwise dry lands necessitated the construction of lengthy irrigation canals and sophisticated flood-control devices and, hence, required a submissive and cheap workforce. Authoritarian bureaucracies emerged to control this workforce, to cajole and discipline it. Despotic structures of governance were required only because the many workers recruited for the arduous task of digging and maintaining irrigation canals and other protective works gained very little from their efforts. The ruling elite had to design a strong bureaucratic apparatus and sophisticated methods of governance to control people and, thus, to ensure maximal water use.[19] Laws had to be promulgated to provide the authority for the bureaucratic activity and for disciplinary measures. Hammurabi's Code, for example, prescribes

---

[18] Karl A. Wittfogel, *Oriental Despotism: A Comparative Study of Total Power* (New Haven, CT, Yale University Press, 1957).
[19] Wittfogel, *Oriental Despotism*, p. 109.

penalties for neglecting the maintenance of irrigation ditches in Mesopotamia.[20] As Wittfogel explains:

Having access to sufficient arable land and irrigation water, the hydraulic pioneer society tends to establish statelike forms of public control. Now economic budgeting becomes one-sided and planning bold. New projects are undertaken on an increasingly large scale, and if necessary without concessions to the commoners. The men whom the government mobilized for corvee[21] service may see no reason for a further expansion of the hydraulic system; but the directing group, confident of further advantage, goes ahead nevertheless. Intelligently carried-out, the new enterprises may involve a relatively small additional expense, but they may yield a conspicuously swelling return. Such an encouraging discrepancy obviously provides a great stimulus for further governmental action.[22]

The logic of these hydraulic societies was strikingly different than the logic of the common pool resources system. These societies were based on a vertical power-relationship between the bureaucratic elite and the peasants. The elite's constant drive for further taming of nature proved to be so unsustainable that it led to its demise. While one part of the tale of these societies is a tale of despotism, the other part is therefore a tale of unsustainability, which resulted in their demise. These hydraulic societies declined and ultimately disappeared, in large part due to the salinization of fields by the sediments in the water carried by the lengthy irrigation canals.[23]

While the ascendancy of the modern state was not based on control of water or other natural resources, controlling these resources did, however, provide opportunities to use the state's central powers, whether the legislature, the bureaucracy, or the judiciary, to interfere with local common pool management structures to provide benefits to larger segments of society beyond the closely knit, but often politically weak, communities. Hence, laws were promulgated to regulate use and resolve ensuing conflicts.[24] These laws allocated property rights in water – some

---

[20] Laws of Hammu-Rabi, No. 55, reprinted in G. R. Driver and John C. Miles (eds.) *The Babylonian Laws: Ancient Codes and Laws of the Near East* (2 vols., Oxford, Clarendon Press, 1952–5), vol. II, p. 31.

[21] Corvee labor is temporary but recurring forced labor. Corvee workers were recruited seasonally, usually before the flooding period. On this type of recruitment, see Wittfogel, *Oriental Despotism*, note 18, at pp. 47–8.

[22] *Ibid.*, at p. 109.

[23] Clive Ponting, *A Green History of the World* (London, Sinclair-Stevenson, 1991) at pp. 69–73.

[24] On the history of domestic water law see Ludwik A. Teclaff, *The River Basin in History and Law* (The Hague, Martinus Nijhoff, 1967); Dante A. Caponera, *Principles of Water Law and Administration: National and International* (Rotterdam and Brookfield, VT,

allowing individual ownership, others vesting ownership in the state –
and authorized administrative agencies and judges to allocate rights and
obligations.

The ascendancy of the modern state and its use of its central powers did not ensure optimal and sustainable use of its natural resources.
Indeed, the story of Sumer and Assyria was often repeated by central
governments set on providing food to their mushrooming populations
and eager to stride towards development. Between 1950 and 1980 there
was an almost threefold increase in the total area of irrigated global
agriculture. This increase augmented agricultural output by between
50 and 60 percent, but at a dear price: many of the large-scale irrigation projects proved to be heavily subsidized and unsustainable economically (when comparing the rate of return to opportunity costs of
capital).[25]

The opportunity to progress through interference with nature, so
tempting for the Mesopotamian rulers, has proved attractive to many,
if not most, governments in the developing world. During the second
half of the twentieth century many of those governments embarked on
water-related mega-projects, such as high dams or extensive irrigation
systems.[26] The national command and control systems they set up for
the management of those projects were fraught with all of the regular maladies of central management. Losses often occurred, allocations
were often skewed, and deprivations were often the result of human
action rather than nature's curse.

Sometimes losses resulted from a poor understanding of hydrology
or of environmental processes. It is believed that poor understanding of
the harsh effects of water sedimentation and of field salinization was
responsible for the demise of the great empires of Assyria and Sumer.
Similarly, the popularity of high dams in the crucial span of about half

A. A. Balkema, 1992). On the development of water law in the United States see
Morton J. Horwitz, *The Transformation of American Law 1780–1860* (Cambridge, MA,
Harvard University Press, 1977), pp. 34–53.

[25] Elinor Ostrom, *Crafting Institutions for Self-Governing Irrigation Systems* (San Francisco, ICS
Press, 1992), pp. 1–7. See also *World Development Report 1992* (*Development and the
Environment*, The World Bank), p. 100: about 73 percent of water withdrawals are
allotted for irrigation and are heavily subsidized. In India, irrigation accounts for 93
percent of water consumption: Salman M. A. Salman, *The Legal Framework for Water
Users' Associations: A Comparative Study* (Washington, DC, World Bank, 1997), pp. 1–2.

[26] See William M. Adams, *Wasting the Rain: Rivers, People and Planning in Africa* (London,
Earthscan, 1992); Fred Pearce, *The Damned: Rivers, Dams, and the Coming World Water
Crisis* (London, Bodley Head, 1992); Patrick McCully, *Silenced Rivers: The Ecology and
Politics of Large Dams* (London and Atlantic Highlands, NJ, Zed Books, 1996).

a century, beginning in the 1930s and ending in the 1980s, is now believed to have contributed to the deteriorating environmental conditions in many countries.[27] Deforestation is now understood to be a major cause of soil erosion and hence of deadly floods and mud slides, including, for example, the 1998 floods that affected one hundred and eighty million people (4,150 of them killed) in the Yangtze basin, and the mud slides that killed hundreds in Italy and exacerbated the deadly results of Hurricane Mitch in Central America.[28] It is argued further that the soil erosion caused by deforestation also increases the amount of silt in rivers that adds salt to fresh water and surface land. Deforestation is also believed to increase heat and reduce rainfall.[29] Deforestation and its consequences are cited as additional causes for the demise of Sumer.[30]

But routinely, losses and skewed decisions emanate not from ignorance or poor judgment, but from the willful burdening of domestic groups by other groups who abuse the inherent flaws that exist in the domestic political processes of states. The management of the environment and water harbors ample opportunities for diverse rent seekers to capture disproportional shares or impose externalities upon others. These opportunities exist in democratic and non-democratic systems alike. Democratic governments often pursue short-term goals to benefit pressure groups or impress voters. Non-democratic governments have used pretentious water projects such as giant dams and large-scale reclamation and irrigation projects to boost productivity and energy for short-term gains to support failing state-run economies. Such projects have enabled rulers to consolidate control over the population in problematic areas and to promote "nation-building" or the people's identification with the leaders.[31] Communist governments, struggling for survival,

---

[27] Edward Goldsmith and Nicholas Hildyard, *The Social and Environmental Effects of Large Dams* (San Francisco, Sierra Club Books, 1984), p. 17.

[28] See the 1999 World Disasters Report (published by the International Federation of Red Cross and Red Crescent Societies).

[29] Norman Myers, "The Anatomy of Environmental Action: The Case of Tropical Deforestation" in Andrew Hurrell and Benedict Kingsbury (eds.), *The International Politics of the Environment: Actors, Interests, and Institutions* (Oxford, Oxford University Press, 1992), pp. 430, 436–7. See also Ponting, *A Green History*, note 23, at p. 258. Deforestation is believed to be responsible for the occurrence of devastating natural disasters.

[30] Ponting, *A Green History*, note 23, at p. 70.

[31] See Adams, *Wasting the Rain*, note 26, and Pearce, *The Damned*, note 26, for two well-researched accounts of the follies, mismanagement, consequent waste of the precious resource, and misery to the millions whose villages and cultures were either inundated by man-made lakes, dried up by diverted streams, or turned salty by ill-conceived irrigation schemes.

have largely ignored the future at the expense of the environment, causing serious ecological disasters as they opt for immediate gains in industry and agriculture.[32] Developing as well as developed states have depleted or seriously overcharged their aquifers, disregarding or under-estimating the demands of future generations.[33] Most governments have tended to play down the interests of minorities, especially indigenous groups. Ultimately, many of the projects approved by all types of govern-ment have proved to be economically inefficient, hydrologically waste-ful, and environmentally disastrous. Several rivers and lakes stand today as dying monuments to short-term national policies. One particularly re-vealing example is the Aral Sea, which, in the last forty years, has shrunk to about 34 percent of its original volume and 60 percent of its origi-nal area, mainly due to communist-era emphasis on growing the highly water-intensive cotton. Storms of salt, dust, and pesticide residues have turned this lake into an environmental hazard.[34]

Disrespect for human rights and minority cultures has also been strongly associated with national management of water and other natu-ral resources. With decreasing supplies and increasing demands, water policy has become a potent and sometimes punishing tool in domestic politics. The water weapon has become a major source of human rights violations, sometimes on a very large scale. In the last few decades, the damming of rivers has caused the uprooting of millions of people, whose dwellings, fertile farmlands, forests, and wildlife have been flooded and culture shattered. Their potential plight was not a major considera-tion in deciding where to erect dams.[35] In the early 1980s, it was es-timated that approximately 16 million people, most of them belonging

---

[32] See Francis W. Carter and David Turnock (eds.), *Environmental Problems in Eastern Europe* (London Routledge, 1993), for the particular problems of post-communist Europe.

[33] On the mining of confined aquifers in Libya and Saudi Arabia, see John Kolars, "The Course of Water in the Arab Middle East" (1990) 33 *American Arab Affairs* 57, 63. On the mining of aquifers by individuals in the western states of the United States, see Edgar S. Bagley, "Water Rights Law and Public Policies Relating to Ground Water 'Mining' in the South-western States" (1961) 4 J. Law & Econ. 144.

[34] On the environmental disaster of the Aral Sea and the efforts to rehabilitate it, see Laurence Boisson de Chazournes, "Elements of a Legal Strategy for Managing International Watercourses: The Aral Sea Basin" in Salman M. A. Salman and Laurence Boisson de Chazournes (eds.), *International Watercourses: Enhancing Cooperation and Managing Conflict, Proceedings of a World Bank Seminar* (Washington, DC, World Bank, 1998), p. 65.

[35] See M. Kassas, "Environmental Aspects of Water Resources Development" in Asit K. Biswas (ed.), *Water Management for Arid Lands in Developing Countries* (Oxford, Pergamon Press, 1980), p. 67; Goldsmith, *Effects of Large Dams*, note 27, at p. 17; Adams, *Wasting the Rain*, note 26, at p. 132.

to minority groups, had become refugees of dam construction.[36] An estimate from 1996 gives a range of 30 to 60 million people, most of them in China and India.[37] Continuing dam construction throughout the world will add significantly to this number. The construction of the Three Gorges Dam on China's Yangtze River alone is expected to cause the forced dislocation of between one and two million people.[38] The Yacyreta dam on the Parana River, expected to become the world's largest, will require, according to official estimates, the relocation of about 8,300 families.[39] Many resettlement schemes have failed to rehabilitate the evacuated people, who have suffered great physiological, psychological, and socio-cultural stress. In several places, indigenous and other minority communities were severely affected due to the central government's wish to increase the availability of water to other segments of the population or to add hydroelectricity. In Brazil, hydroelectric projects that flooded Indian lands affected thirty-four tribes. The Pakistani government forced in 1963 about 100,000 people of the Chakma tribe from their homes in the Chittagong Hill Tracts (in what is today Bangladesh) in order to clear the way for the Kaptai Dam and the settlement of 400,000 Bengali farmers. The Quebec government resettled Native Americans to clear an area of 11,000 square miles of Cree lands for the erection of a dam.[40] The administrative decision-making process hardly involved the adversely affected communities,[41] and the courts are less than eager to protect their interests.[42]

---

[36] Pearce, *The Damned*, note 26, at pp. 154–5.

[37] See McCully, *Silenced Rivers*, note 26, at pp. 66–7 and pp. 321–33.

[38] Audrey R. Topping, "Ecological Roulette: Damning the Yangtze" *Foreign Affairs* (Fall 1995) (the dam is expected to result in the resettlement of 1.4 million people and vast ecological dangers but on the other hand facilitate shipping and provide electricity and flood control).

[39] This dam is a joint Argentinian–Paraguayan project. See *International Environment Reporter* (Bureau of National Affairs, 30 Sept. 1998, at 965).

[40] For these and other affected communities, see Goldsmith and Hildyard, note 33, at pp. 19–50; Pearce, note 26, at pp. 155–6, 218–24; Phillip Hurst, *Rainforest Politics: Ecological Destruction in South-East Asia* (London and Atlantic Highlands, NJ, Zed Books, 1990), p. 197.

[41] Robert Paine, *Dam a River, Dam a People? Saami (Lapp) Livelihood and the Alta/Kautokeino Hydro-Electric Project and the Norwegian Parliament* (Copenhagen, International Work Group for Indigenous Affairs, 1982) (discussing the effects of a proposed hydropower project on the Saamis in Norway and arguing that these adverse effects were not discussed in the parliamentary debates).

[42] United States federal courts have dismissed suits brought by Indian tribes against water projects that inundated sacred sites and cemeteries: see *Sequoyah v. Tennessee Valley Authority*, 620 F.2d 1159 (6th Cir. 1980); *Badoni v. Higginson*, 638 F.2d 172

In addition to refugees of dam construction, similar hardships have be-
fallen individuals and communities who suffered from the drying up of
floodplains and marshlands, another outcome of river impoundment. In
Africa, for example, regulation of flow through dams or diversion canals
stopped the occurrence of floods and, thus, dried up floodplains and
marshlands that, together, supplied almost half of the irrigated area in
sub-Saharan Africa. This brought destruction to the ambience and, consequently, to the tribal way of life for many communities that depended
on these seasonably flooded areas.[43] At times, governments regard these
side effects as desirable. In Turkey, the South-East Anatolia Project (GAP),
which includes the giant Ataturk Dam and large-scale irrigation projects
in Turkey's eastern region, includes plans to resettle large parts of the
local population. These plans are expected to alleviate the economic
situation of the population, but also to strengthen Turkey's control over
the rebellious Kurds and quell their separatist sentiments.[44] In southern
Sudan, the Jonglei Canal was planned to divert the waters of the White
Nile before they enter the Sudd marshes region, where the people in the
Jonglei area – struggling against the Khartoum government – live on
agriculture and aquaculture.[45] Iraq's government has also used the water
weapon to punish dissident minorities in the South.[46]

Indeed, states' control over national natural resources has been a
major factor for both unsustainability and human maltreatment. As demands for these resources increased, so did the governments' inclination

(10 Cir. 1980). On the often lacking protection of minority rights and interests by
national courts see Eyal Benvenisti, "National Courts and the International Law on
Minority Rights" (1997) 2 ARIEL 1.

[43] For such effects as the displacement of inhabitants, destruction of their tribal way of
life, loss of fertile lands beneath the reservoir, and loss of floodplain agriculture and
fisheries downstream in the estuaries and the near-shore area, see Adams, note 26, at
pp. 68–99, 128–54 (1992); Geoffrey E. Petts, *Impounded Rivers: Perspectives for Ecological
Management* (Chichester, Wiley, 1984), pp. 11–12.

[44] On these Turkish plans see H. J. Skutel, "Turkey's Kurdish Problem" (1988) 17(1)
International Perspectives 22, 24.

[45] The civil war between the Christian South and the Muslim North has stopped all work
on the canal for the time being. On the Jonglei Canal Project, see Robert O. Collins,
The Waters of the Nile: Hydropolitics and the Jonglei Canal, 1900–1988 (Oxford, Clarendon
Press, 1990), pp. 311–405; John Waterbury, *Hydropolitics of the Nile Valley* (Syracuse, NY,
Syracuse University Press, 1979), p. 77; Paul Howell, Michael Lock and Stephen Cobb
(eds.), *The Jonglei Canal: Impact and Opportunity* (Cambridge, Cambridge University Press,
1988), especially Part II.

[46] See John Bulloch and Adel Darwish, *Water Wars: Coming Conflicts in the Middle East*
(London, Victor Gollancz, 1993), pp. 137–8. On the life and culture of the marsh
people in southern Iraq, see Wilfred Thesiger, *The Marsh Arabs* (London, Longmans,
1964).

to use them as a domestic political tool. During the twentieth century, this tool has also become a tool of international politics.

Internationalizing the management of natural resources

With the advent of technology and population growth, more and more resources – fresh water, clean air, fisheries, hydrocarbon and mineral deposits, forests, nature reserves, and endangered species of flora and fauna – have become both increasingly scarce and subject, or potentially subject, to national control and management. Because many, if not most, of these resources defy political borders, governmentally managed scarcity has become during the twentieth century a major international concern. Domestic mismanagement has increased the dependency of countries on "international" resources. A dwindling domestic lake increased one riparian's reliance on an international river; the cutting or burning of forests within one state impacted global warming and caused mud slides in downstream states. Nations faced most starkly the ancient choice between conflict and cooperation, as both technology and demands made possible and desirable the regulation of transboundary resources on the international plane through specific agreements and the development of international norms.

The increased demand for transboundary resources has been coupled with an "internationalization" of resources as a result of the breaking up of political boundaries. More resources have become transboundary resources as a consequence of the enormous political changes, from the breaking up of empires after World War I, through the decolonization process, to the collapse of the Soviet Union and the dissolution of Yugoslavia. These changes have spurred new international conflicts over the management of shared resources. Claims for "internal self-determination" that included calls for autonomy for groups in managing local resources upon which they relied, also added to the internationalization process.[47]

---

[47] The draft United Nations Declaration on the Rights of Indigenous Peoples, adopted by the Sub-Commission on Prevention of Discrimination and Protection of Minorities, UN Commission on Human Rights, on 26 August 1994, sets out to ensure, *inter alia*, the right of indigenous peoples to maintain and strengthen their relationship with their lands, territories, waters, and other resources (Article 25), to own and manage these resources (Article 26), and to participate in decisions affecting these resources. See also the decision of the Human Rights Committee in the case of *Länsman et al.* v. *Finland*, Communication No. 511/1992, UNDoc. CCPR/C/52/D/511/1992 (1994) (concerning the claim of Saamis to natural resources as covered by Article 27 of the International Covenant on Civil and Political Rights).

The international community took pains to adequately acknowledge this increasingly blurred distinction between the domestic and the international. As late as 1974, the United Nations' Charter of Economic Rights and Duties of States[48] declared that "every State has and shall freely exercise full permanent sovereignty, including possession, use and disposal, over all its wealth, natural resources and economic activities."[49] It was, however, evident already then that the resources most crucial for human subsistence were dependent on more than one state's unilateral action. Therefore, Article 3 of the Charter stipulated that "in the exploitation of natural resources shared by two or more countries, each State must cooperate on the basis of a system of information and prior consultation in order to achieve optimum use of such resources without causing damage to the legitimate interests of others." The clash between the ideology of "full permanent sovereignty" and the recognition of an obligation to prevent sub-optimal and harmful uses of "shared resources," so well exemplified by the superimposition of the two articles of the Charter, proved to be almost irreconcilable. Characteristically, the definition of what amounts to a "shared resource" eluded governmental delegates.[50] This clash is apparent in the subsequent general instruments dealing with the environment, such as the 1972 Stockholm Declaration of the UN Conference on the Human Environment, the 1992 Rio Declaration of the UN Conference on Environment and Development, and the 1992 Convention on Biological Diversity.[51] Such clashes were, of course, not only in theory, and actual conflicts did arise. Especially acute were the skewed relationships in which one state was able to burden its neighbor by, for example, blocking the flow of the shared river or using the unidirectional winds to pollute the neighbor's air and fields. In such upstream–downstream relationships, lower riparians went as far as issuing aggressive threats to protect their share of successive rivers, while upper riparians refused to recognize their duty to share "their" water with their downstream neighbors. Thus, Turkish–Syrian relations have been strained mostly due to their tense upstream–downstream relations as Turkey developed its capacity to cut the flow of the Euphrates through the mammoth dams of the GAP. One of Syria's responses was to provide support and shelter to the Kurdish guerrilla group (PKK) fighting for

---

[48] UNGA 3281(XXIX), 12 December 1974.   [49] *Ibid.* Article 2.

[50] See Nico Schrijver, *Sovereignty over Natural Resources: Balancing Rights and Duties* (Cambridge, Cambridge University Press, 1997), p. 132. On the difficulties involving the definition of transboundary watercourses, see chapter 7.

[51] On the underlying North–South conflict and the "fragile" consensus that led to the adoption of these formulas, see Schrijver, *Sovereignty*, at pp. 122–40.

secession from Turkey.[52] Egypt, the lower riparian of the Nile, issued military threats against its weaker upstream neighbors to prevent them from contemplating any interference with the flow of the river.[53] China, the upper riparian of the Mekong River, refused to negotiate on the use of the river with its downstream neighbors and did not become a member of the 1995 treaty establishing the Mekong River Commission.[54] India, the stronger power in the Indian subcontinent, acted unilaterally both vis-à-vis downstream Bangladesh and upstream Nepal, constructing dams and barrages that adversely affected India's neighbors.[55] Both China and Turkey refused to recognize their duty under international law to cooperate with their downstream neighbors with respect to their shared rivers and voted against the 1997 Convention on the Law of the Non-Navigational Uses of International Watercourses.[56] The blunt statement of Turkey's president, Suleyman Demirel, at the opening ceremony of the Ataturk Dam on 25 July 1992, the crown of the entire GAP project, captures the upstream state's position:

Neither Syria nor Iraq can lay claim to Turkey's rivers any more than Ankara could claim their oil. This is a matter of sovereignty. We have a right to do anything we like. The water resources are Turkey's, the oil resources are theirs. We don't say we share their oil resources, and they can't say they share our water resources.[57]

[52] On these strained relations, see David Kushner, "Conflict and Accommodation in Turkish–Syrian Relations" in Moshe Ma'oz and Avner Yaniv (eds.), *Syria Under Assad: Domestic Constraints and Regional Risks* (London, Croom Helm, 1986), pp. 85, 95–7. See also John Bulloch, note 46, chapters 3 and 5; Natasha Beschorner, *Water and Instability in the Middle East*, Adelphi Paper 273 (London, Brassey's, 1992), pp. 36–9.

[53] Imeru Tamrat, *Constraints and Opportunities for Basin-wide Cooperation in the Nile – A Legal Perspective* in J. A. Allen and Chibli Mallat (eds.), *Water in the Middle East: Legal, Political, and Commercial Implications* (London and New York, I. B. Tauris Publishers, 1995); Beschorner, *Water and Instability*, at pp. 59–60.

[54] Agreement on Cooperation for the Sustainable Development of the Mekong River Basin, 5 April 1995, rep. in 34 ILM 864 (1995). On this agreement and the challenges involved in its implementation see Philip Hirsch, "Natural Resource Conflict and 'National Interest' in Mekong Hydropower Development" (1999) 29 Golden Gate U. L. Rev. 399.

[55] See Surya P. Subedi, "Hydro-Diplomacy in South Asia: The Conclusion of the Mahakali and Ganges River Treaties" (1999) 93 AJIL 953. On the 1996 Ganges treaty and its implementation see Treaty on Sharing of the Ganga/Ganges Waters at Farakka, 12 Dec. 1996, Bangl.–India (1997) 36 ILM 523; Salman M. A. Salman, "Sharing the Ganges Waters between India and Bangladesh: An Analysis of the 1996 Treaty" in Salman and Boisson de Chazournes, *International Watercourses*, note 34.

[56] On this agreement, see Chapter 7.

[57] Quoted in "Thirsting for War," BBC World Service Friday 29 September 2000 (http://news.bbc.co.uk/hi/english/audiovideo/programmes/correspondent/newsid_946000/946916.stm).

But even in cases where states were mutually dependent, as in the case of border rivers and lakes – when all sides have an equal opportunity to affect the flow's quantity and quality – conflict is often not avoided. The colossal environmental damage to the Rhine is but one glaring example of prevalent collective failure to promote shared interests.[58] In a number of cases water has become political leverage, a powerful weapon to elicit political concessions, and, at least on one occasion, a significant step on the road to military conflict. It was reported in 1998 that Syria had expelled the leader of the PKK guerrilla group, Abdullah Ocalan, in the hope that Turkey would finally agree to negotiate the allocation of the Euphrates River.[59] It is widely perceived that the June 1967 war in the Middle East erupted as a direct consequence of heated border skirmishes along the Israeli–Syrian border following Syrian attempts to divert the headwaters of the Jordan River.[60]

The question of management of transboundary resources has, thus, become a complex issue of international law. This body of law is no longer merely called upon to instruct riparians how to allocate quantities of a shared river or lake. Rather, the reality of deep interdependency between riparians imposes a heavy burden on international law to supply mechanisms for joint management of shared resources. So far, however, this necessary conclusion – that the ideal of permanent and full sovereignty over resources is destructive and that transboundary resources must be treated as shared and regulated by norms beyond each state's unilateral measures – has eluded governments that refused to acknowledge any diminution of their powers. Those states had little incentive to share control, and the international political and legal system failed to change this incentive structure. The international system of sovereign states as a whole failed to rise to the challenge of collective action to manage shared transboundary resources.

The objectives of this book

This book may be viewed as an effort to outline conditions for collective action in the management of transboundary resources. It draws

---

[58] See discussion in chapter 3, notes 26–8 and accompanying text. On the environmental problems of the shared fresh water of Finland and Sweden, see chapter 3, notes 30–1 and accompanying text.

[59] "Turkey Rejects Syrian Demand for Water Talks," Agence France Presse, 25 August 1998 (reprinted in Lexis-Nexis files).

[60] See Beschorner, *Water and Instability*, note 52, at p. 21.

inspiration from the many examples of efficient and sustainable management of common pool resources by small closely knit communities and asks what explains the parallel failure of states to share transboundary resources and develop closely knit communities of states to jointly appropriate them. The effort to respond to this question leads me to inquire whether there are inherent difficulties in the nature of the state that render it an actor less prone to cooperation than the individual actor. By exploring the roots of this relative failure of states to cooperate, this book points out systemic failures in the way states behave. The identification of these failures leads me to propose ways to overcome these failures. Once these failures are addressed, the argument goes, collective action will emerge endogenously, without need for recourse to international norms and procedures that would impose negative incentives upon recalcitrant states.

Underlying this quest is the realization that in the basically anarchic system of international law and politics, in which norms still depend on state consent, there is little prospect for agreement on a global scale with regard to international norms restricting national sovereignty over the exploitation of "national resources." And indeed, as already hinted in this chapter and as will be further demonstrated in chapter 7, state practice in this context is often mixed, and conflicts of interest are reflected in contradictory approaches to the legal regulation of the relevant issues. The book therefore does not aim at identifying the few areas of general agreement and the more areas of disagreement, but instead to provide a basis for analyzing the preferred legal approach to the subject. The book argues that once an "environment of cooperation" exists, cooperation will emerge endogenously. Therefore, the pertinent question is what are the necessary elements of such an environment? Here, again, the insights of the theory and practice of collective action assist in providing a general course of study: if states fail to cooperate because of their heterogeneous nature, because governments do not internalize all the consequences of their actions, then procedures that ensure such internalization could increase the governments' inclination to cooperate. The book therefore must explore in what ways the incentive structure of states differs from that of individuals and whether there are any possibilities of addressing this difference.

In exploring the potential for collective state action, this book breaks away from the usual distinction between domestic and international processes. The systemic failures of states that derive from their heterogeneity in their management of natural resources affect similarly domestic and international resources. Domestic conflicts of interest – between farmers and city dwellers, between producers and consumers – are responsible for distorted governmental choices, whether on the domestic or the international level. In fact, many domestic interest groups cooperate with *foreign* interest groups in order to impose their externalities on their respective rival *domestic* groups. The better-organized and, hence, more politically effective domestic interest groups – usually producers, employers, and service-suppliers – cooperate with their counterparts in different states to exploit collectively the less organized groups in those states, such as the consumers, the employees, and the environmentally affected citizens. Thus, many global collective action failures must be attributed to conflicts between warring domestic groups rather than to international competition.

Hence solutions must be found both in the domestic and the international components of the governmental decision-making process. Therefore, this book adopts a view of the management of shared natural resources as a transnational challenge. It rejects the still-prevailing Westphalian perception of nation-states as unitary actors engaging in international competition over international resources. Instead, it adopts a different paradigm – the transnational conflict paradigm – that explains better various collective action failures and provides guidance with regard to feasible mechanisms for correcting these failures. At its core lies the observation that states are not monolithic entities and many of the pervasive conflicts of interest are in fact more internal than external and stem from the heterogeneity within states.

The transnational conflict paradigm exposes the fact that crucial constitutional and international norms have been designed to perpetuate the unbalanced power relationship between rival interest groups. Current norms and procedures, both constitutional and international, are inherently slanted in favor of groups with historically stronger domestic political power. This institutionalized imbalance was unheeded in international legal scholarship, perhaps due to the traditional division between constitutional law and international law.[61]

Given the prevalence of skewed decision making, mismanagement and conflict, on the one hand, and glimpses of sustainable cooperation on the other, the questions to be explored in this book will address

[61] A similar departmentalization occurred in political science, with the emergence of international relations as a separate branch. See Helen V. Milner, "Rationalizing Politics: The Emerging Synthesis of International, American, and Comparative Politics" (1998) 52 Int'l Org. 759, 762–7.

the issue of transboundary resource management from the combined focus of efficiency, democracy, and equity. The book will seek to identify the collective failures and offer remedies to accommodate these three goals. Identifying mismanagement and waste as resulting from both domestic and international politics, the book will seek a comprehensive approach, linking international law with domestic law. Realizing that scarcity and vulnerability of ecosystem management requires ongoing decision making regarding a host of questions, ranging from the types of plants grown, to the location of industries and power plants, and finally to the water-recreational options available to the general public and the tourist industry, this inquiry will emphasize the possibilities for shared regimes of collective management of transboundary resources.

Collective action in the utilization of transboundary resources can, in principle, provide optimal and sustainable results. A bleak future of wars over control of water resources is not an unavoidable tragedy in our new millennium. Despite ominous predictions of global warming and population explosion, the problem in most cases is not insufficient supplies, but regulating the conflicting demands. Thus, for example, the world still has enough water to meet the existing and future needs of the world's population, at least in the twenty-first century.[62] Both experience and theory tell us that collective action will be the choice if, and only if, the net benefits it provides for all actors are greater than the benefits of individual action. Can we design an international community in which international cooperation will yield higher gains for the individual actors? Are there sufficient incentives for national actors to invest in designing such a community? These are the questions this book sets out to explore.

---

[62] Petts, *Impounded Rivers*, note 43, at p. 1; Peter H. Gleick, "An Introduction to Global Fresh Water Issues" in Peter H. Gleick (ed.), *Water in Crisis: A Guide to the World's Fresh Water Resources* (New York, Oxford University Press, 1993), pp. 3–4.

2 The need for collective action in the management of transboundary resources

Sovereignty, political borders and transboundary commons: the ingredients of tragedies

The two basic building blocks of the global political and legal environment – the concept of sovereignty and the allocation of jurisdiction by political borders – have joined forces to preclude an efficient and sustainable use of transboundary resources ever since the global environmental crisis emerged towards the middle of the twentieth century. By that time, the idea of national sovereignty had reached its zenith. Decolonization and other secessionist movements were vying for equal sovereign status and the unfettered discretion it promised for the management of national resources.

The concept of sovereignty implied a legal environment of unencumbered national control over the resources found within national jurisdiction. The international legal environment was based on what may be called "the Lotus principle," namely, the underlying freedom of states to do whatever is not proven to be prohibited under international law. In other words, a world in which "Restrictions upon the independence of States cannot be presumed."[1] The Lotus principle, enunciated by the Permanent Court of International Justice in 1927 provided that sovereigns were free to dispose of the resources under their jurisdiction at their pleasure, unless a contrary international norm were proved. Until such a norm is proved, a polluter has to be bought off to curb emissions. A heavy water user would have to be similarly remunerated for forgoing

[1] *The Lotus Case* (*France* v. *Turkey*), PCIJ Reports, Series A, No. 10 (1927) (ruling that the claimant state has the burden to prove the existence of an international norm that constrains the action of the defendant state).

its use. The Lotus principle, then, could be regarded as epitomizing an optimistic belief in the international invisible hand.

Sovereignty in itself would not have precluded an efficient allocation of resources among states. The Coase theorem suggests that absent transaction costs, the allocation of resources through trade will be efficient.[2] What precluded efficient outcomes was the fact that sovereignty was allocated to entities whose political borders disregarded the boundaries of the natural resources in question. These natural resources straddled political boundaries that did not take into consideration environmental and hydrological characteristics. At best, transboundary resources such as lakes and rivers were used as natural fences, tools designed in earlier days for keeping neighbors apart and at peace, tools that have become obsolete and even detrimental to cooperation in an age of conflicting demands. At their worst, political borders were drawn with the colonizers' blissful ignorance of geographic conditions in faraway continents. As a result, sovereigns had the opportunity to export waste and degradation to their neighbors, often with impunity. By the middle of the twentieth century, when environmental disasters were becoming increasingly apparent, there was very little room for redrawing political borders to conform to the boundaries of natural resources. Even the newly established former colonies adhered to the principle of *uti possidetis juris*, sanctifying ancient colonial maps drawn with little respect for natural conditions and drainage basins.

Only in a handful of cases were efforts successfully made to adjust the political or natural borders so as to subject the resource to unilateral appropriation and thus to enable its efficient management. Thus, during the establishment of the borders of Palestine in the 1920s, efforts were made to include within its borders the headwaters of the Jordan River. These efforts proved only partially successful.[3] The 1960 Indus treaty between India and Pakistan redrew the natural boundaries by separating

---

[2] Ronald H. Coase, "The Problem of Social Cost" (1960) 3 J. Law & Econ. 1; for the application of the Coase theorem in international settings see John A. C. Conybeare, "International Organization and the Theory of Property Rights" (1980) 34 Int'l Org. 307.

[3] On the 1906 delimitation of the Egypt/Palestine border and the strategic importance of springs in that arid area, see the documents collected in Patricia Toye (ed.), *Palestine Boundaries 1833–1947* (4 vols., Slough, Archive Editions, 1989), vol. IV, p. 602. On the (partially successful) British efforts in 1918–1923 to include the headwaters of the Jordan River within the territory of Palestine see Toye, *Palestine Boundaries*, vol. II, p. 3; H. F. Frischwasser-Ra'anan, *The Frontiers of a Nation: A Re-examination of the Forces which Created the Palestine Mandate and Determined its Territorial Shape* (London, Batchworth

the six main tributaries of the Indus River and assigning the three eastern rivers to India and the three western ones to Pakistan.[4] Also during the 1960s, Israel and Jordan reached a similar informal arrangement concerning the Jordan River, through massive investments in diversion canals, that enabled Israel to use the upper part of the river and Jordan to use most of the Yarmuk River flowing into the lower Jordan.[5] But aside from these three exceptions, control over resources did not form a consideration in the redrawing of political boundaries. Most transboundary resources remained subject to the control of more than one state. As a result, the respective entitlements of states concerning their shares of transboundary resources remained rather vaguely defined. The international invisible hand could not function under such circumstances.

And so, the market-leaning international legal environment converged with the vaguely defined property rights of individual states to produce somber results. Often, military power determined the outcome. Many "transactions" resulted from extortion forced upon the potential polluter or user rather than being freely agreed upon, as the weaker state had to give up its right of use without appropriate compensation. Absent legal constraints and well-defined rights, relatively powerful downstream states began resorting to economic and military threats to elicit cooperation, while relatively weaker riparians acquiesced, and actual practice reflected the regional military balance of power.[6] On a few occasions, states financed pollution abatement measures in other

Press, 1955), pp. 97–141; Adam Garfinkle, *War, Water, and Negotiation in the Middle East: The Case of the Palestine–Syria Border, 1918–1923* (Tel Aviv, Tel Aviv University, Moshe Dayan Center for Middle Eastern and African Studies, 1994), p. 115.

[4] "The Indus Water Treaty 1960" (1960–1) 1 *Indian Journal of International Law*, 341; Richard R. Baxter, "The Indus Basin" in Albert H. Garretson, Robert D. Hayton, and Cecil J. Olmstead (eds.), *The Law of International Drainage Basins* (Dobbs Ferry, NY, published for the Institute of International Law, New York University School of Law [by] Oceana Publications, 1967), pp. 443, 460–1; Aloys A. Michel, *The Indus Rivers: A Study of the Effects of Partition* (New Haven, Yale University Press, 1967), pp. 254–65; Marc Wolfrom, *L'utilisation à des fins autres que la navigation des eaux des fleuves, lacs et canaux internationaux* (Paris, A. Pedone, 1964), pp. 106–18.

[5] This arrangement dates back to the proposals included in the Johnston Plan, which were never officially accepted by the riparians. On the Johnston Plan and its background see, e.g., Kathryn B. Doherty, "Jordan Waters Conflict" in 1965 *International Conciliation*, No. 553 (New York, Carnegie Endowment for International Peace, 1965), pp. 25–8; Georgina G. Stevens, *Jordan River Partition* (Stanford, CA, Hoover Institution on War, Revolution, and Peace, Stanford University, 1965). The solution finally adopted in the Israel–Jordan Treaty of Peace of 26 October 1994, (1995) 324 ILM 46 (Annex II) basically follows the said plan, but it also sets up a Joint Water Committee and envisions joint projects on the Yarmuk and the lower Jordan.

[6] See chapter 1, notes 52–9 and accompanying text.

states that refrained from curbing their own pollution. Thus, the United States constructed and financed waste-water collection and treatment projects along the Mexican bank of the Rio Grande;[7] the Dutch government paid France a sum of about 45 million French francs as its contribution to the effort to reduce the chloride pollution of the Rhine from French territory;[8] and more recently ten developed countries agreed to pay Russia to enable it to meet its obligations under the Montreal Protocol.[9] But these are the exceptions that prove the rule. In general, arguments over entitlements and obligations precluded many agreements that would have promoted optimal use of shared resources.

This chapter explores why the two building blocks of the international legal and political system – sovereignty and political borders that cut through transboundary resources – preclude optimal use of transboundary resources through the invisible hand of the Lotus. It seeks an alternative to the Lotus principle in the context of transboundary resources, and examines whether such an alternative can emerge through collective efforts. This examination yields bad news with respect to the management of global resources, but potentially good news for the management of transboundary resources.

Nature, markets, and the state system

For transnational markets in natural resources to flourish and provide efficient outcomes, the individual entitlements of states sharing them should be neatly defined. If entitlements are not protected against incursion by others, their value is diminished and trade in them will not capture their full worth.[10] But nature does not provide easily and cheaply defined individual entitlements. The choice, therefore, is between two possibilities: to invest in defining individual entitlements or to forgo such differentiation and develop alternatives to market transactions. Such a choice is dependent on the estimated costs of differentiation of

---

[7] David J. Eaton and David Hurlbut, *Challenges in the Binational Management of Water Resources in the Rio Grande/Rio Bravo* (Austin, TX, Lyndon B. Johnson School of Public Affairs, University of Texas at Austin, 1992), pp. 55–77.

[8] Alexandre C. Kiss, "Commentaire" (1986) 2–3 Rev. Jur. Envn't 307; Kiss, "Commentaire", (1983) 4 Rev. Jur. Envn't 353, 354. On the pollution and the litigation that ensued see chapter 3, note 26 and accompanying text.

[9] Agreement from 7 October 1998, rep. in the *BNA Environmental Reporter*, vol. 21, p. 1025 (1998); www.unep.org/unep/secretar/ozone.

[10] See Yoram Barzel, *Economic Analysis of Property Rights* (2nd edn, Cambridge, Cambridge University Press, 1997), pp. 5–6.

entitlements and of the transaction costs of trade in them. The higher the definition and transaction costs involved in the protection and transfer of the entitlements, the lower the likelihood of an optimal use of the resources, and hence the higher the prospects of a better outcome for an alternative system.[11]

To understand why a clear definition of entitlements of property rights in parts of transboundary resources does not make sense, we must turn to their characteristics and their complex interactions. All actions related to the use of air, water, and land may affect the environment in one way or another. The opportunities for externalizing the costs of any social activity – farming, use of certain grains and pesticides, choice of energy sources, emissions, sewage release – are manifold. In the absence of a reliable framework to address such externalities, users' property rights are poorly defined. Even if we were to assign individual entitlements to parts of a shared resource, say a land area, the possibilities of neighbors adversely affecting its value through emissions, sewage release, or diversion of water sources would hamper efficient trade in the land.[12] As many environmentalists have observed, even after initial entitlements in natural resources have been defined in law or practice, there is no escaping regulating all potentially environment-affecting action. Moreover, transaction costs are often high among the different users and affected communities, who are burdened by a lack of monitoring capabilities, of competence to assess the harmful effects of activities, and of the ability to prevent free-riding and holdouts.[13]

In international settings, the parceling out of entitlements to environment-affecting behavior is even more difficult. The definition of an entitlement, the definition of its infringement, and the assignment of responsibility for infringement by governments or individuals all depend on a collective effort to define rights and duties and on a system for enforcing them.

It is important to realize that water resources pose a particularly great challenge for a Coasean, market-based regime. Water resources are particularly difficult to define in neat, tradable property rights. As opposed

---

[11] See Barzel, *Economic Analysis*, pp. 126–7.

[12] Daniel C. Esty, "Revitalizing Environmental Federalism" (1996) 95 Mich. L. Rev. 570; Carol M. Rose, "Rethinking Environmental Controls: Management Strategies for Common Resources" 1991 Duke L. J. 1; Peter L. Kahn, "The Politics of Unregulation: Public Choice and Limits on Government" (1990) 75 Cornell L. Rev. 280.

[13] See Guido Calabresi and A. Douglas Melamed, "Property Rules, Liability Rules and Inalienability: One View of the Cathedral" (1972) 85 Harv. L. Rev. 1089, 1215; Esty, "Revitalizing", note 12, at pp. 577–99.

to air, which is still very much in abundance, water poses unique problems of supply, as well as diverse, rather rigid demands. Diverse demands – domestic, agricultural, industrial, recreational – must be met from constantly fluctuating resources. Aside from "confined aquifers" – unreplenished underground pools of "fossil water" that accumulated prehistoric rains – water is in a constant motion of evaporation, precipitation, filtration, and run-off. It is subject to periodic predicted and unpredicted fluctuations as well as disasters that strain significantly its availability. Its quantity and quality influence the condition of other media such as soil and climate and the condition of these other media, in turn, affect the water. Existing resources must be constantly monitored against illegal appropriation and against pollution.

Aquifers pose yet a more difficult challenge in this respect. Property rights in them cannot be simply delineated. In underground systems, water flows usually at a relatively slow rate, as the water must permeate porous rock. Thus, withdrawal effects may take many years to travel from well to well. Without monitoring, it is impossible to verify the amounts of water pumped by the co-riparians. Furthermore, underground reservoirs are sensitive to over-pumping and contamination, which may cause irreversible damage. Sometimes it is difficult to determine whether an aquifer has been polluted or to identify the source of the pollution. Often, the treatment of a polluted aquifer requires swift action. The karstic properties of limestone rocks through which water percolates to the ground cause caves and underground canals to form and create shortcuts for pollutants seeping into the aquifers. Natural impediments like sorption or filtration are absent under these conditions. In light of all these considerations, many issues must be regulated, including the location of wells, monitoring of amounts actually pumped and of water quality, artificial recharge of aquifers, conservation of aquifers, and prevention of pollution.

One possible exception to the definitional problem involves new sources of water, such as reclaimed sewage water for irrigation[14] or desalinated water for domestic uses. The assignment of ownership – to their producers – is not troublesome. Moreover, it makes good sense to

---

[14] Sewage water can be treated even to reach potable standards and certainly could be used for irrigation. Urban sewage usually consists of 55 percent of the fresh water supplied: H. Bouwer, "Reuse of Water: A Sustainable Perspective" in J. C. Van Dam and J. Wessel (eds.), (1993) 1 *Transboundary River Basin Management and Sustainable Development*, 89; World Health Organization, "Health Guidelines for the Use of Wastewater in Agriculture and Aquaculture", Technical Bulletin Series 77; Israel State Comptroller Report on Mekorot Water Company Ltd., 35, (1995), (in Hebrew).

provide the proper incentive to public and private enterprises to create such sources.

Beyond the problem of definition of property rights in transboundary resources and, in particular, shared fresh water, lurks the problem of the high transaction costs involved in the trade in such rights. Such trade, unless strictly regulated, would tend to disregard third party interests and impose externalities on their shares. This is due to the natural interdependency of ecosystems and the multitude of interests involved. Take, for example, a source of water that irrigates a field. A large part of this water percolates to the underground and is then recaptured by nearby users. In fact, the same water is being used and reused copiously all along the course of a river, charging and recharging the underlying aquifer. When riparian A uses a certain amount of water for irrigation, a substantial part of it drains into the ground or returns directly to the river and, ultimately, benefits other riparians. Any transactions in A's water share that entail the transfer of water to users outside the basin or to riparians situated below B along the river will adversely affect B and deprive her of a portion of her share.[15] Therefore, A's sale of any amount to outside users would not entail compensation to the many Bs downstream and, hence, would not reflect the true value of the resource to all the riparians. Moreover, A's transaction would fail to account for other, less identifiable third party interests, such as the loss of recreational opportunities for nature-loving hikers and bathers further downstream. Theirs is a public good not protected by the market.[16]

---

[15] For the effects of return water flows on trade in water rights and suggestions for overcoming these effects, see Terry L. Anderson and Pamela Snyder, *Water Markets: Priming the Invisible Pump* (Washington, DC, Cato Institute, 1997), pp. 92–8 (recognizing the need for administrative processes to account for return flows); H. Stuart Burness and James P. Quirk, "Water Law, Water Transfers, and Economic Efficiency: The Colorado River" (1980) 23 J. Law & Econ. 111; Ronald N. Johnson, Micha Gisser, and Michael Werner, "The Definition of Surface Water Right and Transferability" (1981) 24 J. Law & Econ. 273 (suggesting that water rights be defined by the consumptive use of each riparian: namely, its use minus the return flow); Charles J. Meyers and Richard A. Posner, *Market Transfer of Water Rights: Towards an Improved Market in Water Resources* 290 (National Water Commission, Legal Study No. 4, NTIS No. NWC-L-71-009, July 1971) (cited by Johnson *et al.* at 273–4) (calling for assigning property rights in the return flow as well). For the burgeoning literature on water markets since the 1980s, see Norman J. Dudley, "Water Allocation by Markets, Common Property and Capacity Sharing: Companions or Competitors?" (1992) 32 Nat. Res. J. 757, 758–63.

[16] Gregory A. Thomas, "Conserving Aquatic Biodiversity: A Critical Comparison of Legal Tools for Augmenting Streamflows in California" (1996) 15 Stan. Envtl. L. J. 3 (analyzing the institutional arrangements under federal and California laws to intervene on behalf of the public interest in water trade and demonstrating their shortcomings – mainly the discretionary character of the regulatory powers; suggesting, instead,

Thus, for example, the resistance mounted by the Canadian federal and provincial governments against the private export of water from Canada to the US underlines the concerns of third party losses from unrestricted market exchanges.[17]

The analysis of the natural characteristics of fresh water and other transboundary resources does not necessarily indicate that the allocation of specific entitlements to be traded on markets is impossible. Rather, it means that trade in such individually assigned entitlements requires regulation. When shared natural resources are involved, regulation that encompasses the entire resource in question is mandatory to achieve efficient and sustainable use. All existing domestic water or pollution markets are subject to regulation that defines entitlements and monitors compliance.[18] This conclusion applies with equal force to international settings and is sufficient for rejecting as harmful a market-based policy based on the Lotus "hands-off" approach.

Our immediate conclusion is, therefore, that whatever system we choose to adopt for the management of transboundary resources, be it market-based or a command and control approach, we first have to entrench the system in a regulatory regime that defines initial entitlements and provides the necessary infrastructure for their efficient allocation. Any market-based system will require a strong monitoring and enforcement mechanism to replace the otherwise simple enforcement mechanisms of fences, boundaries, and the like, and to protect third party interests. This is the conclusion of all observers of existing inter-state trading systems that exist in the United States and Australia.[19] This

---

a "market" response – purchase by a central agency of water from users to maintain the flow – as California's Water Code allows since 1992; funds may be procured through taxation.

[17] On this question, see Jamie W. Boyd, "Canada's Position Regarding an Emerging International Fresh Water Market with Respect to the North American Free Trade Agreement" (1999) 5 NAFTA L. & Bus. Rev. Am. 325; Sophie Dufour, "The Legal Impact of the Canada–United States Free Trade Agreement on Canadian Water Exports" (1993) 34(2) Les Cahiers de Droit 705.

[18] On the regulated water market in New Mexico, see Johnson et al., note 15; on the water market in California, see Morris Israel and Jay R. Lund, "Recent California Water Transfers: Implications for Water Management" (1995) 35 Nat. Res. J 1; on the water market in Texas, see Ronald C. Griffin and Fred O. Boadu, "Water Marketing in Texas: Opportunities for Reform" (1992) 32 Nat. Res. J 265. For extensive analysis of water trade in Spain and the United States, see Arthur Maass and Raymond L. Anderson, . . . And the Desert Shall Rejoice: Conflict, Growth and Justice in Arid Environments (Cambridge, MA, MIT Press, 1978).

[19] See Barton H. Thompson, "Water Markets and the Problem of Shifting Paradigms" in Terry L. Anderson and Peter J. Hill (eds.), Water Marketing – The Next Generation (Lanham, MD, Rowman & Littlefield, 1997), pp. 1–22 (suggesting to address environmental

is also the conclusion of a historical analysis of ancient trading systems since biblical times: The trade in water among villagers in Palestine required the formulation of a substantial body of law concerning intra- and inter-group transactions in water shares and responsibilities for maintaining ditches and other waterworks. The result is reflected in intricate Bedouin customs,[20] and norms of Jewish[21] and Muslim law.[22]

In internal basins, national procedures can enforce control of the entitlements, including monitoring actual withdrawals, compensating affected third parties, including the general public, and sanctioning excessive and abusive uses. In international basins, agreements must establish similar arrangements.[23] Such complex arrangements must be based upon intra-basin agreements and institutions, involving the majority, if not all, of the riparian states. In the international setting, there is no escape from collective action. Such collective action must start with the state of nature, namely, the ecosystem with its intricate and complex interaction between water, climate, winds, soil, flora, and fauna in each specific region. Nature implies that riparian states share property. This is the starting point for further collective action.

externalities of out of basin transfer by "remediation of the harms or a tax equal to the costs"); James L. Huffman, "Institutional Constraints on Transboundary Water Marketing" in Anderson and Hill, *Water Marketing*, p. 31, at p. 39 ("The reduction or elimination of [third party losses which are] obstacles to transboundary water markets is dependent on the creation of some form of transboundary institutions."), Gary L. Sturgess, "Transborder Water Trading among the Australian States" in Anderson and Hill, *Water Marketing* pp. 127–45 (describing the Australian interstate water market in the Murray Basin as a response to growing demands, noting "insufficient consideration has been given to environmental end-use and instream requirements" and describing efforts to take these into account, at 137).

[20] Bedouin customs in the Negev and Sinai deserts are described by Captain Owen of the British army in a letter dated 16 July 1906, rep. in *Palestine Boundaries*, note 3, at vol. I, p. 602.

[21] On Jewish law developed in the Talmudic period see Refael Patai, *Ha-Maim (The Water)* (Tel-Aviv, 1936, in Hebrew); Mordechai haCohen, *Halachot veHalichot (Norms and Manners)* (Jerusalem, 1975, in Hebrew), pp. 106–16.

[22] On Moslem law see Chibli Mallat, "Law and the Nile River: Emerging International Rules and the Shari'a" in P. P. Howell and J. A. Allen (eds.), *The Nile, Sharing a Scarce Resource: A Historical and Technical Review of Water Management and of Economic and Legal Issues* (Cambridge, Cambridge University Press, 1994), pp. 365, 372–5; Mallat, "The Quest for Water Use Principles: Reflections on the Shari'a and Custom in the Middle East" in J. A. Allen and C. Mallat (eds.), *Water in the Middle East: Legal, Political and Commercial Implications* (London, I. B. Tauris Publishers, 1995).

[23] See Gerhard Hafner, "The Optimum Utilization Principle and the Non-Navigational Uses of Drainage Basins" (1993) 45 Austrian J. Publ. Int'l Law 113, 134–6.

Therefore, we have to invest in a collective effort to provide a system for the definition of initial entitlements of states sharing transboundary resources. Whether such a collective effort is feasible, whether international and domestic norms can facilitate it, are the questions that must be discussed, and will be discussed in this chapter. Note that at this stage we do not eschew the market-based approach for resolving transboundary resource use and favor command and control mechanisms. At this preliminary juncture we must explore the necessary preconditions – including legal norms – for enabling either the first or the second approach (or a combination of both). We will return to this debate in chapter 6, when we pit the "regulated market" approach against the "command and control" approach.

Fortunately, and somewhat paradoxically, the natural character of shared transboundary resources holds the promise of collective action leading to effective shared regulatory mechanisms. A comprehensive definition of the common property that nature provides can often become a basis for fruitful cooperation among the co-owners. To understand this, we must delve into the study of collective-action theory.

The theory of collective action applied to transboundary resources

Because different states enjoy access to transboundary natural resources, they face a collective action problem. Each state is interested in getting more out of the resource with minimal costs, and these interests conflict with those of the other users. This conflict can lead the parties to a race to the bottom. Garret Hardin's famous tale of "the tragedy of the commons"[24] is the prevailing nightmare that captures this thought. But such a race is not the only possibility. Under certain conditions, to be explored here, that devastating race could be replaced with fruitful and sustainable cooperation. Abundant literature on common pool resources, spawned mainly by Mancur Olson's seminal book on collective action[25] and supported by insights from game theory, has shown that in such a type of resources, cooperation among appropriators can and does occur endogenously when certain conditions are met. Rational egoists are likely to develop procedures for cooperating toward the optimal

---

[24] Garret Hardin, "The Tragedy of the Commons" (1968) 162 *Science* 1243, 1244.

[25] Mancur Olson, *The Logic of Collective Action* (Cambridge, MA, Harvard University Press, 1965). See also Todd Sandler, *Collective Action: Theory and Applications* (Ann Arbor, University of Michigan Press, 1992), pp. 8–9.

use of such resources. This theory implies that instead of investing in parceling the commons by delineating individual entitlements and in maintaining enforcement institutions to back the entitlements up, a rather unlikely prospect, one should first examine how to enhance the internal interaction between the co-owners of the commons and thereby promote efficient cooperation.

To assess the possibility and promise of collective action in the management of transboundary commons, it is necessary to first delve into the characteristics of these resources. In economic literature, there is much discussion regarding the distinction between types of goods, ranging from pure public to pure private goods.[26] Pure public goods consist of goods whose benefits are *nonexcludable* and *nonrival*. They are nonexcludable because it is impossible or prohibitively costly to prevent outsiders from gaining access to them. They are nonrival since a user's consumption of a unit of that good does not detract from the benefits that accrue from the good to others. Thus, for example, the neighbor who cleans the sidewalk provides a public good that each pedestrian enjoys, without diminishing the enjoyment of other pedestrians. In contrast, the benefits of a private good, such as a loaf of bread, are fully excludable and rival. The user may prevent others from using it, and the consumption of any part detracts from the whole. There are two other types of goods positioned between pure private and pure public goods: (1) public goods that are nonexcludable, yet rival, which may be consumed by all who gain access to them, yet their consumption detracts from the consumption of others, such as ocean fisheries or open-access pastures; and (2) common pool resources, which are partially excludable and rival.[27] Transboundary natural resources to which more than one state enjoys access, such as regional ecosystems or hydrocarbon deposits straddling two countries, are examples of common pool resources. The benefits from such resources are partly excludable. In contrast to open-access commons such as fisheries beyond 200 nautical miles of Exclusive Economic Zones, the ocean floor, or the electromagnetic spectrum, only states that enjoy access to the transboundary resource can benefit from it directly. The benefits from these resources are also rival, since any

---

[26] On the different types of goods, see, e.g., Sandler, *Collective Action*, note 25, at pp. 5–7; Elinor Ostrom, *Governing the Commons: the Evolution of Institutions for Collective Action* (Cambridge, Cambridge University Press, 1990), p. 30; Russell Hardin, *Collective Action* (Baltimore, MD, Johns Hopkins University Press, 1982), pp. 17–19; Michael Taylor, *The Possibility of Cooperation* (Cambridge, Cambridge University Press, 1987), pp. 5–6.

[27] Ostrom, *Governing the Commons*, note 26, at pp. 30–3; Hardin, *Collective Action*, note 26 at 19; Taylor, *Possibility of Cooperation*, note 26 at p. 3.

unit of the resource used by one state reduces the amount available to the other co-owners.

Both pure public goods and transboundary resources are susceptible to the "tragedy of the commons" syndrome in which each of the appropriators receives direct benefits from his or her unilateral act, while the costs imposed by these acts are borne by others. There is, however, a very important distinction between common pool resources and pure public goods. With regard to common pool resources, the possibility of excluding outsiders provides an opportunity for the limited number of insiders to coordinate their activities and, thereby, avert a tragedy of their shared commons. In other words, with transboundary resources such as ecosystems, which are international common pool resources, there is a palpable potential for collective action between the co-owner states that will provide an optimal and sustainable use of the resource.

The number of participating actors may influence the tendency to cooperate. When there are more actors, coordination is more difficult to attain.[28] The costs of coordinating actors' activities and monitoring their performance are likely to increase because more formal methods for coordination and monitoring are required. But in the case of transboundary resources, there are only a handful of states sharing each transboundary resource. If each of the participating states can control the activities of its domestic actors, then coordination between a handful of international actors should not be an unmanageable task.

The basis for analyzing collective action issues is game theory. Collective action problems involving transboundary resources can be formalized by different game structures. Prisoner's Dilemma (PD) is the most famous game; but other games such as Assurance, Chicken, and Stag Hunt are also of relevance to our subject, since many of these game situations exist between states sharing a transboundary resource. PD games[29]

---

[28] See Taylor, *Possibility of Cooperation*, note 26 at p. 105; Sandler, *Collective Action*, note 25, at p. 48; Hardin *Collective Action*, note 26, at pp. 182–5; Edna Ullmann-Margalit, *The Emergence of Norms* (Oxford, Clarendon Press, 1977), p. 47.

[29] The game derives its name from the story of two prisoners who are kept in different cells and are interrogated. Each of them is confronted with two courses of action (strategies), between which each must choose simultaneously and without knowing the other's choice. They can either "cooperate," i.e., choose a strategy that will leave both of them better off, or "defect," with each one choosing a strategy that maximizes his own payoff, regardless of the other's loss. In the classic PD game, the strategies are to cooperate by not confessing to committing the offence of armed robbery or to defect by confessing and thus giving testimony against the other player. If they remain silent (cooperate), they will each receive a one-year sentence for illegal possession of weapons, as the police will have no evidence of the robbery; if one confesses (defects)

can be associated, for example, with several issues concerning water utilization and management. Take the example of two riparians who draw water from a shared lake or an aquifer. These riparians can cooperate by keeping withdrawals lower than the replenishment rate and by acting to prevent pollutants from reaching the resource. By cooperating, the two riparians incur certain costs (lower rate of consumption, improvement of infrastructure), but they also ensure sustainable use of the resource. The sustainability of the resource depends on both riparians' cooperation. Without an effective means of communication and enforcement of commitments, however, riparian A cannot be sure whether riparian B will choose to incur the costs and cooperate or to defect and use the resource without limits. In such a situation, the dominant strategy of both riparians will be to defect. By defecting, they reach a Pareto-inferior outcome. As the PD game seems to suggest, our two riparians face the prospect of either depleting the resource or else contaminating it beyond repair. On the basis of the 1993 and 1995 agreements between Israel and the PLO, the two parties face this game with respect to the Mountain Aquifer, the most important water resource for both communities.[30]

Another pervasive game structure in natural resources contexts is the Assurance game.[31] In this game, an actor will cooperate if and only if the other actors cooperate as well, because the actor values cooperation more than unilateral action. The inclination to favor cooperation is often enhanced by external considerations, unrelated to the resource

and the other remains silent (cooperates), the first is pardoned and the latter gets a ten-year sentence; if both confess (defect), they both get five-year sentences. PD is an example of a situation in which two rational actors are driven to choose a Pareto-inferior outcome: faced with the two strategies, each player has a dominant strategy – to defect – no matter what strategy the other player chooses. Even if they agree beforehand to cooperate, neither has an incentive to honor the agreement.

[30] See Israeli–Palestinian Interim Agreement on the West Bank and the Gaza Strip, Washington, 28 September 1995, Annex III, Protocol Concerning Civil Affairs, Appendix 1 – Powers and Responsibilities for Civil Affairs, Article 40 (Water and Sewage), and Schedule B. On this issue, see Eyal Benvenisti and Haim Gvirtzman, "Harnessing International Law to Determine Israeli–Palestinian Water Rights" (1993) 33 Nat. Res. J. 543; Eyal Benvenisti, "The Israeli–Palestinian Declaration of Principles: A Framework for Future Settlement" (1993) 4 Eur. J. Int'l L. 542, 552–4.

[31] On the Assurance game in the environmental and conservation context, see Daphna Lewinsohn-Zamir "Consumer Preferences, Citizen Preferences, and the Provision of Public Goods" (1998) 108 Yale L. J. 377; Christopher H. Schroeder, "Rational Choice Versus Republican Moment Explanations for Environmental Laws, 1969–73" (1998) 9 Duke Env. L. & Pol'y F. 29; Donald T. Hornstein, "The Political Origins of Modern Environmental Law: Self-interest, Politics, and the Environment – A Response to Professor Schroeder" (1998) 9 Duke Env. L. & Pol'y F. 61.

in question. A sense of solidarity or a common cause often reduces the temptation to free-ride on the efforts of others and thus transforms many PD situations to Assurance games. Thus, for example, it is believed that strong social ties facilitated cooperation between farmers in Haiti to prevent soil erosion, despite the fact that a significant minority (36 percent of the actors) lacked any private interest in the value of the protected lands and had no reason other than reputation as cooperators to invest efforts.[32] The same sense of solidarity, achieved through the cultivation of a shared religion, facilitated cooperation among villages in Bali, Indonesia.[33]

The Chicken game is also manifested in many environmental situations.[34] Take the PD game example of the two riparians drawing water from a shared lake or an aquifer, but now assume that the resource is polluted and that each riparian can clean it up unilaterally. Each riparian prefers that the other clean up the resource, while the second-best preference is that both cooperate by cleaning it so that the resource becomes usable again. Other scenarios in the form of the Chicken game occur where one of the two riparians can unilaterally prevent flooding from a shared river (by constructing a dam) or depletion of an aquifer (by unilaterally curbing one's own withdrawals). The one riparian will tend to bear the costs unilaterally if she prefers this outcome to the risk that the other would not do anything and the field would be flooded or the resource ruined.

Finally, the management of transboundary resources can raise scenarios that lend themselves to Stag Hunt situations.[35] Instead of a stag,

---

[32] T. Anderson White and C. Ford Runge, "The Emergence and Evolution of Collective Action: Lessons from Watershed Management in Haiti" (1995) 23 *World Development* 1683.

[33] See chapter 1.

[34] This game derives its name from the notorious game in which two drivers speed at each other from opposite directions. The driver who swerves out of the way (cooperates) first is designated the "Chicken." If neither swerves (i.e., if both defect), the outcome is disastrous for both. Mutual defection has the lowest payoff for both. Each driver's first preference is that the other cooperate, and her second-best preference is that both cooperate. In this case, neither driver has a dominant strategy, and each driver's choice depends on her assessment of the other's probable action. Since each prefers to defect if the other cooperates, both are expected to resort to pre-game acts that would convey to the other player her determination to defect. A Pareto-inferior outcome can result from concurrent irrevocable acts by the two players committing themselves to non-cooperation. However, one would expect at least some players (the more risk-averse ones) to cooperate, be the "chicken," and prevent the worst outcome.

[35] A group of hunters attempt to capture a stag. Every hunter must cooperate in order to capture the stag. All prefer capturing the stag, but if a hare happens to pass by one

our riparians face, for example, the prospect of erecting a dam that will increase water availability to all. From a topographic and hydrologic point of view, the best site for the proposed dam is on the territory of a midstream riparian. The midstream state's cooperation entails the loss of arable and inhabited lands that would be flooded by the dam's waters. Upstream riparians contribute by ensuring unhampered flow of the water downstream. The downstream riparians sacrifice their control of the river flow upon which they have relied from time immemorial. The utilization of the Nile is a case in point: upstream–downstream cooperation in the Nile basin would achieve much better use of the Nile waters. Impounding the Nile in Lake Tana in Ethiopia, for example, could prevent the loss of as much as 15,000 million cubic meters per year of much-needed water — almost a fifth of the Nile's annual flow — from evaporation and seepage from the downstream Lake Nasser, which is created by the High Aswan Dam in Egypt.[36] Egypt's opposition to this plan derives from the consequences of relinquishing its control over the flow: there are no guarantees that once this stag is caught, the benefits would be fairly divided.

It is not necessarily the case that all riparians have the same preferences. For example, riparian A may have PD preferences, while riparian B has Chicken preferences.[37] In this example, riparian B, who values the outcome of cooperation more than riparian A, will "cooperate" unilaterally. In addition, the structure of the game may change over time. For example, a PD game can turn into a Chicken game and

---

of the hunters, he might catch and eat it, thereby depriving the other hunters of the opportunity to capture the stag (but not the hares, if they begin chasing them). This situation can take the form of an Assurance game, which is a coordination game. In this game, players do not have dominant strategies. Each player's strategy depends on the strategies of the others: player A will cooperate if all others cooperate and will defect if all others defect. The same story can take the form of a PD game, if we assume that the individual hunters would be tempted to defect when a hare passed by, especially if they anticipate that the other hunters will do the same and they have no means of monitoring and reprimanding fellow hunters. Hunters would tend to defect, especially in the latter situation, and the Pareto-optimal outcome will not be reached. The hunters may also be deterred by the unknown outcomes of cooperation: how would the benefits of the captured stag be distributed among the hunters?

[36] This loss also increases the salinity of the water downstream: Fred Pearce, *The Damned: Rivers, Doms, and the Coming World Water Crisis* (London, Bodley Head, 1992), p. 116. For potential "stags" in the Nile, see Dale Whittington and Elizabeth McClelland, "Opportunities for Regional and International Cooperation in the Nile Basin" (1992) 17 *Water International* 144.

[37] Taylor, *Possibility of Cooperation*, note 26, at pp. 39–40.

vice versa. Overexploiting a resource due to a PD situation may hasten its extinction. At this stage, because the payoffs for overexploitation are highly negative, the dominant strategy of either riparian may become cooperation.[38] When an aquifer dwindles or when one or a number of riparians become more dependent on a resource, their preferences will shift from PD preferences to Chicken preferences or even to those of the Assurance game. Indeed, the growing awareness of the need for international collaboration in the utilization of transboundary resources reflects the changing game preferences of many countries.

The games described above illuminate the risks involved in decision making under conditions of uncertainty. Rational actors may choose strategies that would produce Pareto-inferior outcomes.[39] Does this mean that external intervention is a prerequisite to reducing uncertainty and enforcing agreements between riparians to cooperate? Not necessarily. For our purposes, the important lesson to be gleaned from the abundant literature on collective action is that cooperation may emerge despite the self-interest of the actors, provided that the game is iterated indefinitely.

Game theory demonstrates that indefinitely[40] iterated PD games between the same players are likely to induce them to cooperate, even without external intervention. In iterated games, each player's payoff is the cumulative payoffs of the expected sequence of games. If the number of iterations is indefinite and if the "shadow of the future" is high enough, namely, the players assign a sufficiently high value to the expected payoffs from future iterations of the game, then each player is expected to choose the strategy of conditional cooperation, or "friendly tit for tat." By using the implicit threat of retaliation against defection by the other, the players can elicit cooperation. The same "tit for tat" strategy will produce cooperation also in a PD game played by a group

---

[38] See Sandler, *Collective Action*, note 25, at pp. 122–3.

[39] An outcome is Pareto-optimal if there is no other outcome by which one actor will be better off without someone else being worse off. Stated differently, a Pareto-optimal outcome occurs when there is no other outcome preferred by one actor that is at least as good for the other actors. Any outcome that is not Pareto-optimal is Pareto-inferior.

[40] If both players know beforehand the number of games to be played, the dilemma remains. This is because the dilemma will exist in the final round, at which point, both will defect. With the outcome of that game known, the one-before-last game effectively becomes the final game, and the same argument applies to it, and so on, back to the first game: David M. Kreps *et al.*, "Rational Cooperation in the Finitely Repeated Prisoners' Dilemma" (1982) 27 J. Econ. Theo. 245.

larger than two players and even when some of the players choose to defect unconditionally.[41] Thus, PD situations, which are the least likely to produce cooperation when played only once, are in fact potentially cooperative games, provided the actors remain in the game for an indefinite period of time.

Cooperation based on indefinitely iterated games is to be expected not only in PD games. Such is certainly the case with Assurance games, where cooperation is the highest-ranking preference.[42] Even the indefinitely recurring Chicken game, the game deemed less susceptible to recurring cooperation,[43] can forge cooperation between actors.[44] It may be too risky to act as Chicken, when the recurring game involves issues of national security and total annihilation is a possible outcome. But the stakes are quite different when the game involves the management of shared resources. In this context, the benefits of iterated cooperation, even unilaterally, may outweigh – usually for the more risk-averse players – the costs of defection.[45]

Indefinite iteration is, therefore, an important condition for the emergence of cooperation in the management of transboundary resources. To encourage cooperation, arrangements must be made to ensure that the

---

[41] Michael Taylor, note 26, chapter 4. In the latter example, the strategy of the cooperating players will be to cooperate if a sufficient number of others cooperate.

[42] See Daphna Lewinshon-Zamir, "Consumer Preferences," p. 377.

[43] The deterrence theory in international relations suggests, for example, that the more the Chicken game is expected to recur, the more each player would invest in establishing a reputation for defection. See, e.g., Thomas C. Scheling, *Arms and Influence* (New Haven, Yale University Press, 1966), p. 55: "the main reason we [Americans] are committed in many places is that our threats are interdependent. Essentially we tell the Soviets that we have to react here because if we did not, they would not believe us when we say we will react there."

[44] This is especially true in the case of N-Person iterated Chicken games. In N-person Chicken settings, some of the actors can provide the collective good by cooperating. Two of three riparians can, for example, curb their withdrawals from a shared resource and prevent its ruination, while the third actor free-rides on their act. When it is necessary that at least some (S) of the N group of actors cooperate, each actor must choose whether to join the cooperating group or to defect and enjoy the free ride. If the actor assumes that the cooperating group will be smaller than S–1, she will tend to defect, since she will assume that despite her cooperation, the collective good will not be provided. By establishing a reputation for defection, actors reduce the probability of the other actors cooperating, and ultimately the collective good will not be provided. For an analysis of this claim, see Hugh Ward, "The Risks of a Reputation for Toughness: Strategy in Public Goods Provision Problems Modelled by Chicken Supergames" (1987) 17 Brit. J. Pol. Sci. 23.

[45] Ward, "Risks of a Reputation" at 23, 31. Actual cases that prove this point are the unilateral efforts of downstream states to incur expenses unilaterally to curb pollution by the upstream states: see notes 6–8 and accompanying text.

actors' game continues indefinitely and that exchanges are frequent. Riparians in an interdependent situation may hesitate to cooperate if through cooperation, the game structure might change and they would no longer be able to reciprocate the defection of others. The key to successful cooperation in utilization of transboundary resources is, therefore, the maintenance of indefinite interdependence between the riparians.

The logic in practice

The theoretical analysis outlined above is supported by numerous case studies that describe how efficient and sustainable cooperation in the management of shared fresh water and other transboundary resources emerged among individual actors endogenously. We can now refer back to the examples of endogenous cooperation between villagers described in chapter 1 and explain the root of their success by the theory of collective action. The villagers in ancient Persia and Palestine and in contemporary Bali and India understood the futility of allocating individual property rights in the local spring or small river. Instead, the indefinite iteration of their investment and allocation games created strong incentives for compliance, incentives that led to the thickening and strengthening of their interdependency through social, cultural, and even religious ties. These ties proved flexible to changes in the demand and supply of the resource and persisted much longer than the externally imposed arrangements prescribed and enforced by the "Oriental despots." The smaller villagers proved more resilient and sustainable than the mighty empires of Assyria, Sumer, and Maya. The village communities survived in diverse, often difficult and changing environments, thanks to highly complex systems for joint decision making, effective monitoring and enforcement mechanisms, and the provision of ample information to users regarding water availability and distribution.[46] The lack of a strong central government and formal law enforcement procedures did not prevent successful cooperation and efficient water utilization among

---

[46] Elinor Ostrom analyzed these examples of efficient and sustainable management of natural resources and examples of inefficient ones (Ostrom, *Governing the Commons*, note 26); see also Elinor Ostrom, *Crafting Institutions for Self-Governing Irrigation Systems* (San Francisco, ICS Press, 1992); Shui Yan Tang, *Institutions and Collective Action: Self-Governance in Irrigation* (San Francisco, ICS Press, 1992) (analyzing the factors contributing to success and failure in a wide array of irrigation communities and discussing social differences and the zero-sum honor game (*izzart* to Pakistani farmers) as negative factors that may preclude cooperation [at pp. 68–73]).

these "close-knit groups,"[47] groups in which "informal power is broadly distributed among group members and the information pertinent to informal control circulates easily among them."[48]

These indigenous examples demonstrate the strength of the theory. The most important insight that this theory and practice offers is that the indivisibility of transboundary resources into smaller units does not preclude cooperation. In fact, just the opposite happens: the very interdependency among appropriators of a shared resource can provide strong incentives for enough members to cooperate. These ancient examples can offer inspiration for contemporary designs that will be based on mutuality rather than on coercion. Such designs need not emulate the villagers' old practices. Contemporary legal tools can be used to form the contemporary locus of activities of the newly formed closely knit groups. Contemporary company law could be used effectively. As several writers have shown,[49] private organizations have emerged as satisfactory providers of a solution to water utilization in these areas. Barton Thompson analyzed the functioning of water institutions, most of them private, in seventeen western states in the United States.[50] These institutions, comprising both water suppliers and water users, bring water from their sources to the various users. The two groups, suppliers and users, reduce the transaction costs involved in internal bargaining over prices and quotas by employing internal procedures that include common decision making and provision of ample information concerning water quantity and quality to the users. These procedures provide internal monitoring and enforcement mechanisms and eliminate the need to rely on the external legal system. Ultimately, these internal mechanisms enable stable cooperation and result in an optimal utilization of the water. To enable collective action between states to emerge,

---

[47] See Robert C. Ellickson, *Order without Law: How Neighbors Settle Disputes* (Cambridge, MA, Harvard University Press, 1991). See also Ellickson, "A Hypothesis of Wealth-Maximizing Norms: Evidence from the Whaling Industry" (1989) 5 J. Law Econ. & Org. 83.

[48] Ellickson, *Order without Law*, note 47, at pp. 177–8. The villagers in biblical Palestine maintained their close-knit societies by restricting the allocation only to owners of lands surrounding the village, typically all belonging to a few *hammulahs* or extended families who lived in those villages (see chapter 1, notes 2–6).

[49] For a description of incidents of successful collective action leading to optimal utilization of water resources in a number of basins in Spain and the US, see Maass and Anderson, *And the Desert*, note 18. See also the discussion in chapter 1, text to notes 1–17 and Ostrom, *Governing the Commons*, note 26, chapter 3.

[50] Barton H. Thompson, Jr., "Institutional Perspectives on Water Policy and Markets" (1993) 81 Calif. L. Rev. 671, 687–93.

international law could be used to offer equivalent legal tools to structure similar interdependent institutions.

Regional institutions for collective action: conclusions for an international agenda

The insights of the collective-action theory and practice suggest a more refined approach to international efforts to prescribe norms with respect to allocation and management of transboundary resources. These insights call for an inquiry into the interesting contrast between the villagers' impressive successes and the international failures: how can we reconcile the numerous examples of efficient and sustainable local regimes for common pool resources management with the dismal status of transboundary use and of international law, after almost a century of prescriptive attempts? Understanding the differences between individual farmers and neighboring states as participants in a collective action dilemma may explain why states fail so miserably in promoting collective action endogenously.

Moreover, the analysis of the challenges of collective action can point out, for those states wishing to promote cooperation, the necessary framework for cooperation. In a sense, the theory is a blueprint for the legal engineers who would be called upon, once the political will is forged, to design the institutions for cooperation.

Finally, the theory suggests that international law should become sensitive to the common pool phenomenon, as distinct from the challenges posed by the pure public goods, or, in our context, the "global goods," such as the global climate or ocean cleanliness. Global goods problems are difficult to overcome because the inability to exclude outsiders invites free-riding and increases the costs of monitoring behavior. Moreover, there can be conflicting incentives among the relevant states due to different stages of economic development or other circumstances.[51] Global warming, for example, creates divergent incentives for different states because some of them actually stand to gain from the expected warming and moistening of their region, whereas others will suffer a disproportionate loss due to desertification. Their diverging incentives

---

[51] Conflicting incentives can be expected to exist between developed and developing countries with respect to the desirability of, for example, antitrust laws and high labor standards. See Andrew T. Guzman, "Is International Antitrust Possible?" (1998) 73 NYU L. Rev. 1501 (discussing antitrust laws); Katherine Van Wezel Stone, "Labor and the Global Economy: Four Approaches to Transnational Labor Regulation" (1995) 16 Mich. J. Int'l. L. 987, 996–7 (discussing labor regulation).

may preclude an endogenous provision of the public good in the international sphere.[52] As distinct from the treatment of global commons, international law should not seek to regulate transboundary resource management externally. Instead, global efforts must be made to reduce the costs of endogenous interaction between the resource users. As demands for fresh water and the environment intensify and diversify, interdependency becomes greater, as does the incentive to cooperate in formulating entitlements and norms and in managing shared resources. This progression is consistent with the observation of economists on the emergence of a clearer delineation of property rights as a result of the increasing value of the assets owned.[53] Thus, in the near future, as supplies of finite transboundary resources in general fail to meet rising demands, more and more states may be unable to afford the luxury of ensuring their relative power gap through unilateral appropriation. They will sooner or later opt for making the most from the collective-action situation. As Gary Libecap explains, the incentive of actors to agree upon the definition and allocation of property rights in a shared asset is shaped not only by the magnitude of the common pool losses, but also by the nature of the process of defining them, the costs involved, and the costs of enforcing the defined rights.[54] A body of norms designed to lower the costs involved with collective action will hasten the occurrence of such action. This analysis suggests that the global efforts to define entitlements to transboundary resources must be based on a careful examination of the nature of the transboundary resources in question and the challenges involved in their management. It is certainly not conducive to use the traditional examination of past state practice and prescribe norms for contemporary challenges based on a look back into history.[55]

---

[52] Todd Sandler, *Global Challenges: An Approach to Environmental, Political, and Economic Problems* (Cambridge, Cambridge University Press, 1997), pp. 112–13.

[53] Barzel, *Economic Analysis*, note 10, chapter 6.

[54] Gary D. Libecap, *Contracting for Property Rights* (Cambridge, Cambridge University Press, 1989), pp. 12–14.

[55] See chapter 7.

3 States as collective actors

What motivates states?

The theory and observed practice of collective action provide us with some hope with respect to the possible evolution of cooperation between states sharing transboundary resources. As availability becomes scarce, the net benefits from investments in shared institutions become positive. But such optimism may be unwarranted, in light of the special character of states as collective actors in the international arena. The shadow of sovereignty casts a serious doubt over the potential of interstate cooperation. This chapter explores the nature of states as collective actors. Its goal is to explain why we too often see a failure on the part of states to act in accordance with the predictions on the theory of collective action. The chapter deals first with the doubt regarding the tendency of states to seek relative rather than absolute gains. Unlike individual actors, states often do not seek to maximize their respective gains from their transactions, but, rather, prefer to increase the power gap between themselves and other states. The chapter then delves into the more complex question of the nature of the state as a heterogeneous rather than a unitary actor. Unlike individual farmers who share a local spring or commons, the argument goes, states have also internal conflicts of interests, and hence their behavior and constraints at the international game level may differ from those of individual actors. Do these differences mete out a devastating blow to the predictions based on the theory of international collective action?

Absolute or relative gains?

One complicating factor to the analysis of states as actors is the concern regarding the implicit assumption of the theory that actors seek absolute gains from cooperation. The contrary observation, said to be supported by empirical evidence, is that states do not seek to maximize their respective gains from their transactions, but, rather, prefer to increase the power gap between themselves and other states. This interest in relative welfare is familiar on the individual level, as human beings tend to compare themselves with their fellows. As Thomas Hobbes observed, peoples' joy consists of comparing among them.[1] Experiments by psychologists have shown that on both the individual level and the collective level, people and communities tend to adopt strategies that widen the gap between themselves and the "others," rather than strategies that promote absolute gains.[2] The interaction between ethnic groups often demonstrates a similar inclination.[3] This interest in relativity is accentuated by scholars of the neo-realist school of international relations theory who maintain that considerations of relative rather than absolute power play an even more prominent role in fashioning interactions between states. Because states are preoccupied with the threat of annihilation, the scholars claim, states do not seek to increase their absolute welfare, but, rather, to secure their survival through increasing or maintaining their relative welfare. In the words of Kenneth Waltz, in the anarchic system of international politics "relative gain is more important than absolute gain."[4] As Joseph Grieco argues, "relative capabilities . . . are the ultimate basis for state security and independence in the self-help context of international anarchy."[5] Each state may be content with less than an optimal share, as long as its neighbor receives even less. The neo-realists argue that since states focus on relative gains, they are engaged in a zero-sum game, and in this game, interstate cooperation is unlikely to emerge.

[1] See Thomas Hobbes, *Leviathan* (W. G. Pogson Smith ed., Oxford, Clarendon Press, 1909): "Vertue generally, in all sorts of subjects, is something that is valued for eminence; and consisteth in comparison. For if all things were equally in all men, nothing would be prized" (at 52). "Man, whose Joy consisteth in comparing himself with other men, can relish nothing but what is eminent" (at 130).

[2] Henry Tajfel, "Experiments in Intergroup Discrimination" (1970) 223 *Scientific American* 96. Several experiments with Prisoner's Dilemma and other game situations showed that players defect and incur losses, if only to prevent their competitors from improving their lot: Morton D. Davis, *Game Theory: A Nontechnical Introduction* (revised edn, New York, Basic Books, 1983), pp. 129–33, 151–3.

[3] Donald L. Horowitz, *Ethnic Groups in Conflict* (Berkeley, University of California Press, 1985), pp. 143–7.

[4] Kenneth Waltz, *Man, The State and War: A Theoretical Analysis* (New York, Columbia University Press, 1959), p. 198

[5] Joseph M. Grieco, *Cooperation among Nations: Europe, America, and Non-Tariff Barriers to Trade* (Ithaca, NY, Cornell University Press, 1990), p. 39

Even if one accepts this observation as reflecting states' preferences, one cannot derive from it the neo-realists' conclusion: cooperation does not necessarily entail changing the relative power positions. States seeking maintenance of their relative power positions may agree on arrangements that would ensure that the accumulated gain from cooperation be allocated in proportion to the states' relative power positions[6] or otherwise provide side-payments in various ways to those participants who, through cooperation, gain less than their initial relative positions.[7] This suggestion is supported by a game-theoretic analysis of the relativity argument. Michael Taylor shows that a game played by players who seek to increase their relative gains (what he calls the "difference game") is basically a Prisoner's Dilemma game. In this game, every player's dominant strategy is not to contribute to the collective effort rather than to contribute unilaterally and benefit also the other, free-riding and hence relatively better-off actors. Even Chicken situations that are more problematic to the initiation of cooperation are transformed to PD games when relativity becomes the dominant strategy. Therefore, relativity does not hinder the emergence of cooperation when transactions between parties are repeated indefinitely.[8]

But the observation that states seek to increase their relative power is too crude and should not be too easily accepted. Some states, mainly liberal democracies who share a sense of solidarity, seek to increase the absolute welfare of their citizens and are less preoccupied with interstate comparisons. Therefore, in their relations with similar states, they seek absolute rather than relative gains.[9] The success of the United Nations Economic Commission for Europe (ECE) in the area of freshwater use may be a good example of fruitful cooperation based on strong political and social ties. The ECE has been very effective in coordinating activities, initiating treaties and setting up institutional frameworks throughout Europe concerning water and the environment.[10] Even beyond the

---

[6] This point is implied by Stephen D. Krasner in "Global Communications and National Power: Life on the Pareto Frontier" (1991) 43 *World Politics* 336.

[7] For examples of such compensation arrangements, see Grieco, *Cooperation Among Nations*, note 5, at p. 233.

[8] See Michael Taylor, *The Possibility of Cooperation* (Cambridge, Cambridge University Press, 1987), p. 117.

[9] On the interrelationship between liberal states, see Anne-Marie Burley, "Toward an Age of Liberal Nations" (1992) 33 Harv. Int'l L. J. 393; Burley, "Law among Liberal States: Liberal Internationalism and the Act of State Doctrine" (1992) 92 Colum. L. Rev. 1907 at 1914–28.

[10] For a description of its past activities see Economic Commission of Europe, *Two Decades of Co-operation on Water* (1988). The ECE also initiated the 1992 Helsinki Convention on

liberal realm, relativity may not always reign. In general, states' attitudes towards absolute or relative gains depend upon the constraints they are faced with.[11] Even if neighboring states are preoccupied with their relative power positions, they sometimes cannot afford to forgo cooperation in the management of certain precious transboundary resources. If both riparians have a strong interest in the utilization of a shared resource, with no room for strategic behavior, they are most likely to opt for cooperation. In such a predicament each riparian will consider its absolute rather than relative gain from cooperation. In the near future, as demands for transboundary resources grow and supplies dwindle, more and more states may not be able to afford the luxury of ensuring their relative power gap through unilateral use of shared resources. A body of norms designed to lower the costs involved with such cooperation will accelerate the occurrence of cooperation.[12]

International or transnational competition?

The second concern involves not the dominant strategies of the actors but their very character. The theory of collective action assumes unitary actors who can be made to internalize fully the costs of their actions. The observed cases of successful endogenous cooperation in common pool resources involve actions of individuals only. Had the Westphalian vision of states as unitary actors been true, many problems of transboundary resources management would have been quite successfully resolved. States, as individuals, would have shared the incentive to ensure long-term enjoyment from the transboundary resource. But states are different from individual actors. It is the discrepancy between the Westphalian paradigm and reality that explains many past and current

---

the Protection and Use of Transboundary Watercourses and International Lakes, reprinted in (1992) 31 ILM 1312 and the 1992 Helsinki Convention of Transboundary Effects of Industrial Accidents, reprinted in (1992) 31 ILM 1333.

[11] See Robert Powell, "Absolute and Relative Gains in International Relations Theory" (1991) 84(4) Am. Pol. Sci. Rev. 1303 (arguing that the different assumptions of realists and neo-liberal institutionalists are not incompatible, but, in fact, may be reconciled). See also Robert O. Keohane and Lisa L. Martin, "The Promise of Institutionalist Theory" (1995) 20 Int'l Sec. 39, at 44–6.

[12] Gary D. Libecap, *Contracting for Property Rights* (New York, Cambridge University Press, 1989), pp. 12–14 : the incentive of parties to agree upon the definition and allocation of property rights in a shared asset is shaped not only by the magnitude of the common pool losses, but also by the nature of the process of defining them, the costs involved, and the costs of enforcing the defined rights.

tragedies.[13] Although the Westphalian paradigm treats states as unitary actors in the international sphere, states are in fact heterogeneous. The governments that represent states must shape their policies while juggling with an array of domestic, often conflicting, interests. These conflicting domestic interests and constraints shape the states' international choices.

Why does heterogeneity militate against cooperation? For collective action to sustain, the involved actors must share an interest in an indefinite reiteration of their cooperative moves. This interest exists only when the actors give sufficiently high value to their gains in the indefinite future. In other words, they must have a sufficiently large "shadow of the future."[14] Here lies the major obstacle to cooperation among heterogeneous rather than unitary actors. Transboundary ecosystems constitute resources that could be exploited for diverse purposes. The many conflicting demands over such resources provide ample room for fierce domestic competition involved in their appropriation. Current domestic conflicts over ecosystems involve different uses, from supplying drinking water and sanitation, through sustaining agriculture, industries and recreation, to sewage disposal. They involve diverse users not all equally endowed with the political clout necessary for staking their claims in domestic institutions. The general population within each state generally has a keen interest in long-term cooperation that would ensure long-term availability and inter-generational equity. But other domestic actors, namely, bureaucrats, politicians, and especially the business sector (farmers and industries of various sorts), tend to discount the long-term national benefits of cooperation and prefer partisan short-term goals. While domestic politicians and bureaucrats pursue their short-term goals, their long-term externalities are borne by the general population (including, of course, future generations) and, in particular, by the underrepresented indigenous peoples that live in the affected region.

Domestic political constraints spill over to the international scene, as governments compete to ensure a larger share of the transnational resource to satisfy domestic pressures rather than implement at home politically painful demand management policies. They fear that such painful policies would be resisted by influential domestic interest groups, or could prompt domestic investors to relocate their activities

[13] Eyal Benvenisti, "Exit and Voice in the Age of Globalization" (1999) 98 Mich. L. Rev. 167.
[14] Robert Axelrod, *The Evolution of Cooperation* (New York, Basic Books, 1984), p. 126.

to other jurisdictions that offer competitive conditions.[15] The observed practice of domestic interest groups is to use their political clout to shape their government's position vis-à-vis neighboring states. These groups seek their own sectoral gains, rather than the collective gains of all citizens. In fact, it is often the case that domestic interest groups co-operate with similarly situated *foreign* interest groups in order to impose externalities on their rival *domestic* groups. The better-organized and, hence, more politically effective domestic producers, for example, would cooperate with foreign producers to exploit together the less organized groups in their respective countries, such as employees or environmentally vulnerable citizens. Those groups of producers would use their influence on their respective governments as a vehicle for that purpose. This observation supports a new paradigm, in lieu of the Westphalian paradigm – that explains better the sources of competition over transboundary resources. This is the transnational conflict paradigm.

The rest of this chapter is devoted to exploring the transnational conflict paradigm. It demonstrates first the influence of the domestic politics on the formation of national policies. It identifies those domestic political forces and actors who have a larger say in the domestic legal and political processes, and those who are the disenfranchised groups. Finally, this chapter assesses the negative ramifications of this domestic conflict.

Shifting paradigms: from Westphalia to the transnational conflict paradigm

Ample evidence indicates that domestic groups influence the strategies adopted by "the state" on the international level.[16] But international law, as well as critical analyses of it by lawyers and political scientists,

---

[15] See Daniel C. Esty "Revitalizing Environmental Federalism" (1996) 95 Mich. L. Rev. 570.

[16] For literature on this subject, see Kurt T. Gaubatz, "Democratic States and Commitment in International Relations" (1996) 50 Int'l Org. 108; Robert D. Putnam, "Diplomacy and Domestic Politics: The Logic of Two-Level Games" (1988) 42 Int'l Org. 427. For a discussion of possible sources of internal political influence upon democratic states' attitudes towards cooperation in water-related issues, see David G. LeMarquand, *International Rivers: The Politics of Cooperation* (Vancouver, Westwater Research Centre, University of British Colombia, 1977), pp. 15–19 . He identifies three sources: national bureaucracy, political leadership, and private interest groups. The third source is cited as having been responsible for collective-action failures in international fisheries: M. J. Peterson, "International Fisheries Management" in Peter M. Haas, Robert O. Keohane, and Marc A. Levy (eds.), *Institutions for the Earth* (Cambridge, MA, MIT Press, 1993), p. 249 at 258.

has so far refrained from addressing this crucial phenomenon. The international scene is still characterized by the Westphalian paradigm, namely, the model that views global conflicts as international conflicts among the two hundred some sovereign states that constitute the primary building blocks of the global arena. This paradigm is premised on the still-prevailing perception of nation-states as unitary actors engaging in international competition. Collective-action failures are explained and solutions are suggested on the basis of this paradigm.[17]

The transnational conflict paradigm suggests that states are in fact composed of many competing domestic groups, and hence the competition among these groups is reflected in the external policies adopted by the state. This paradigm explains better various collective-action failures and provides guidance to feasible mechanisms to correct these failures. It provides the necessary basis for understanding the dynamics of scarce resource competition in the international context and, hence, for providing normative and institutional responses to market failures. By looking through the veil of sovereignty, the transnational conflict paradigm explains better the sources and cures of international conflicts and agreements. We suddenly discover that conflicts do not necessarily spring from interstate disagreements on allocation of benefits and costs. Rather, in many cases, conflicts stem from transnational competition among rival domestic groups or even from collusion between several interest groups, all in an effort to capture a disproportionately larger share and externalize costs at the expense of other interest groups within those states, including future generations.

The Israeli–Jordanian water situation can be illustrative of the first point. In May 1997, two and a half years after the signing of the peace treaty between the two countries and one and a half years after the due date under the treaty, the Israeli government finally agreed to transfer a portion of water from the Jordan River to Jordan.[18] This was in fact

---

[17] See, e.g., Andrew T. Guzman, "Is International Antitrust Possible?" (1998) 73 NYU L. Rev. 1501 (reaching a pessimistic conclusion due to conflicting national policies which result from conflicting interests of producers and importers in different states). Other writers continue to impute interests and expectations to states, despite awareness of unexplored domestic dimensions. See Jeffrey L. Dunoff and Joel P. Trachtman, "Economic Analysis of International Law" (1999) 24 Yale J. Int'l L. 1 at 20–1. A notable exception is Schwartz and Sykes' analysis of domestic influences on states' attitudes towards free trade: Warren F. Schwartz and Alan O. Sykes, "The Economics of the Most Favored Nation Clause" in Jagdeep S. Bhandari and Alan O. Sykes (eds.), *Economic Dimensions in International Law* (New York, Cambridge University Press, 1997), p. 43.

[18] The details on the agreement were reported to the Israeli government and public only two weeks after its conclusion: "Israel will Desalinate for Jordan 50 Million Cubic

a forced transfer of revenues from one group of Israeli farmers, located in the Beit-Shean Valley in Israel, to another group, the Jordanian landowners on Jordan's side of the valley. Due to this transfer, these landowners could continue to irrigate their lands without yielding their relatively large share to the thirsty domestic users in Jordan's cities. The Israeli farmers lost out because they had less leverage on the Likud government than under the former Labor-led government (which, in fact, had balked at making the same concessions and stalled the negotiations for months). At the same time, the domestic pressure exerted by the Jordanian landowners on the king was more effective: these landowners, a very influential interest group in Jordan,[19] sought their dividend for the support they had given to the controversial peace with Israel. The Israeli farmers had to satisfy themselves with non-binding promises for future side-payments.[20] This forced exchange was clearly an outcome of negotiations between two governments seeking to further the partisan interests of domestic groups, in which the government that was more domestically vulnerable at the particular time won.[21] No attempt at long-term planning was made. The populations in the Jordanian cities continued to suffer from an insufficient supply of water the following summer.[22]

Meters of Water per Year from the Springs of the Galilee Lake and the Gilbo'a Mountains," *Ha'aretz*, 25 May 1997 (Israeli daily, in Hebrew). Apparently, the agreement comprises a letter from Israel's National Infrastructure Minister, Mr. Ariel Sharon, to Jordan's Crown Prince Hassan of 20 May 1997, based on the Hussein–Netanyahu meeting on 8 May 1997: Zeev Schiff, "An Awkward Agreement with Jordan," *Ha'aretz*, 13 June 1997 (in Hebrew).

[19] On the political influence of the landowners see Rami G. Khouri, *The Jordan Valley: Life and Society Below Sea Level* (London, Longmans, 1981), pp. 192–3 ; Toby Dodge and Tariq Tell, "Peace and the Politics of Water in Jordan" in J. A. Allan (ed.), *Water, Peace, and the Middle East* (London, Tauris Academic Studies, 1996), p. 169 at p. 182; Manuel Schiffler, "Sustainable Development of Water Resources in Jordan: Ecological and Economic Aspects in a Long-Term Perspective" in J. A. Allan and Chibli Mallat (eds.), *Water In The Middle East: Legal, Political and Commercial Implications* (London, I. B. Tauris Publishers, 1995), p. 239 at pp. 243–5.

[20] Consisting of subsidies for research of potential intensification of uses of the reduced quotas. Schiff, "An Awkward Agreement," note 18,

[21] The Israeli farmers are quite influential. They control the water institutions. As a result, they obtain high water quotas at a heavily subsidized price. These policies impose "severe economic damage" to the national economy: see Israel State Comptroller, *Report on the Management of the Water Economy in Israel* 49–50 (1990).

[22] "Jordan's Water Crisis Unresolved," *United Press International*, 3 August 1998. This crisis involved also a political crisis amid allegations that Amman was receiving contaminated water from the Lake of Galilee. This clearly could not be the case, since the lake's waters were being used contemporaneously in Israel without any health or other problems.

Due to domestic constraints, governments tend to address short-ages by adopting strategies that increase supplies, including from internationally shared sources, rather than embark on politically ar-duous demand-management strategies, although such strategies should be the first priority in terms of sustainability.[23] Often governments find it opportune to respond to their domestic public's demand for a larger share, not by allocating less to the powerful small groups, but by present-ing a tougher stance on the international scene, thereby mobilizing pub-lic opinion against the neighboring states. Again, the Israeli–Jordanian context is a telling case in point. During February 1999, it became clear that the coming year would be one of the driest on record. If there were to be sufficient water for the population, deep and painful cuts in the amount supplied to the agricultural sectors in both states would have had to be made. Decision-makers in both governments faced tough political choices: to what extent could they reduce the supply to irri-gated agriculture? To what extent could they wrest more from their neighbor? Israel was in a better strategic position: because the Lake of Galilee, within Israel's territory, served as the storage area for the shared waters, Israel controlled the levers that divert water to the Jordanian pipes. At the same time, the incumbent government was facing a tough challenge in the upcoming elections in May, when every vote could be de-cisive. Not surprisingly, Israeli officials opted to put pressure on Jordan, hoping it would concede. For Jordan, however, water scarcity was high on the political agenda. All still had fresh memories from the previous summer, during which severe failures in the distribution systems had provided foul water or none at all to the residents of the main towns, resulting in strict rationing programs and the sacking of the Water and Irrigation Minister.[24] The rationing system offered little relief for the domestic users. More cuts were politically impossible. Public apprehen-sion did not leave room for the young King Abdullah, only weeks after

---

[23] See J. A. Allan, "Overall Perspectives on Countries and Regions" in Peter Rogers and Peter Lydon (eds.), *Water in the Arab World: Perspectives and Prognoses* (Cambridge, MA, Harvard University Press, 1994), p. 5 at pp. 95–100 (describing the unmet challenges of the Arab states in the Jordan catchment area to engage in effective demand management).

[24] During the long drought, the *Jordan Times* carried daily reports on the water situation, including the rationing program (see "Tighten Your Taps: Water Ministry Announces Summer Rationing Programme", *Jordan Times*, 12 May 1999, available at http://www.access2arabia.com), and imposing quotas on influential landlords in the Jordan Basin (see "Water Plan to be Presented Soon: Government to Set Controls on Wells, Domestic Supplies," *Jordan Times*, 17 March 1999).

his father's death and too weak to face the strong landowners, to make concessions to Israel. Ultimately, the agreement, as well as the farmers' quotas, survived intact,[25] but at the price of increased pumping from overexploited resources.

Transnational collusion of interest groups can be demonstrated by numerous examples relating to international watercourses. Thus, for example, the governments of the upper-riparian states tolerated the continuous pollution of the Rhine River throughout the 1970s and 1980s out of deference to those industries that treated the river as their private backyard dumping area. Little has been done to protect those downstream damaged by the pollution, and suits brought by Dutch environmentally conscious NGOs failed to provide a remedy.[26] For many years, France refused to implement an agreement to reduce the pollution of the Rhine with chlorides due to strong local resistance in Alsace.[27] Insufficient precautions were taken to avert accidental spills, such as the disastrous Sandoz spill in 1986.[28] Yielding to the interests of the domestic industries and their workforces, the governments of the upper-riparian states colluded to stall effective plans to reclaim and protect the river by sustaining their support for a largely ineffective international protection system.[29] Similarly, Finland and Sweden, whose cooperation is usually

[25] In Israel, the cuts for the agriculture sector reached only a 2–5 percent level (20–50 MCMY, based on different estimates): *Ha'aretz*, 3 December 1999, p. C2.

[26] For background on the litigation in Dutch courts with respect to pollution of the Rhine by a French mining company in Alsace, see *Handelskwekerij G. J. Bier BV and Stichting "Reinwater"* v. *Mines de Potasse d'Alsace SA (MDPA)* (1980) 11 Netherlands Yb. Int'l L. 326; for subsequent litigation in this dispute, see *Handelskwekerij G. J. Bier BV and Stichting "Reinwater"* v. *Mines de Potasse d'Alsace SA (MDPA)*, (1984) 15 Netherlands Yb. Int'l L., 471, and *Mines de Potasse d'Alsace SA (MDPA)/Onroerend Goed Maatschappij* (1988) 19 Netherlands Yb. Int'l L., 496. For similar litigation in the French courts, see *La Province de la Hollande septentrionale contre Etat Ministre de l'Environnement*, Tribunal Administratif de Strasbourg, 27 July 1983, reprinted in (1983) 4 Rev. Jur. Envn't 343 (annulling the permit to pollute), overturned Conseil d'Etat, 18 April 1986, reprinted in (1986) 2–3 Rev. Jur. Envn't 296 (finding no manifest error in granting the permit). For a discussion of an "unsatisfactory" 1994 decision of another local court in The Netherlands concerning pollution by a Belgian company of the Meuse River, see Jan M. van Dunne, "Liability in Tort for the Detrimental Use of Fresh Water Resources under Dutch Law in Domestic and International Cases" in Edward H. P. Brans *et al.* (eds.), *The Scarcity of Water* (London, Kluwer Law International, 1997), pp. 196, 205.

[27] Alexandre C. Kiss, "Commentaire" (1983) 4 Rev. Jur. Envn't 353 at 354.

[28] See Aaron Schwabach, "The Sandoz Spill: The Failure of International Law to Protect the Rhine from Pollution" (1989) 16 Ecology L.Q. 443 (discussing the consequences of the toxic waste contamination of the Rhine from a warehouse in Basel and examining the failure of the Rhine treaty regime).

[29] Schwabach points to the interests of the polluting industries and their labor forces as the determining political factors: "The Sandoz Spill," note 28, at 269–71. In its

cause for envy as well as emulation,[30] failed to agree on jointly reducing pollution from pulp mills due to the successful resistance to regulation initiatives mounted by their respective pulp industries, which refused to internalize their costs of production.[31] There are numerous other examples of the ways in which collusion and capture on the part of interest groups affect international negotiations and agreements.[32]

Looking through state actors: who shapes the state's external agenda?

Political economists long ago demonstrated convincingly that state institutions constitute an effective tool for better-endowed domestic interest groups to exploit their competing domestic groups. This theory of public choice views the political system as a market for trade in political goods such as taxes, subsidies, and market regulation.[33] It is common to refer to these exchanges as the result of either concessions by incumbent officials in return for campaign contributions or so-called "pork-barrel" transactions among members of the legislature. In this market, small groups, namely, those composed of a relatively smaller number of individuals, may outbid larger groups, because the smaller the group, the higher the per-capita benefit from cooperation with fellow group members, and the lower the costs of monitoring and sanctioning against free-riding.[34] As a result – excluding discrete and insular ethnic, national, racial, and indigenous minorities[35] – the political influence of

judgment against the Alsatian potash company, the Dutch Court of Appeal determined that the level of pollution allowed by the 1976 Bonn Salt Convention was nevertheless unacceptable under Dutch tort law (see note 26, the 1988 decision).

[30] For a discussion on the very effective Swedish–Finnish Frontiers River Commission, whose tasks are to approve works affecting the flow and quality of the shared water resources and to adjudicate disputes among individual users, see Malgosia Fitzmaurice, "The Finnish–Swedish Frontier Rivers Commission" (1992) 5 Hague YB. Int'l L. 33 at 44.

[31] See Matthew R. Auer, "Domestic Politics and Environmental Diplomacy: A Case from the Baltic Sea Region" (1998) 11 Georgetown Int'l Envtl L. Rev. 77.

[32] For further discussion, see the next chapter.

[33] See the seminal article by George J. Stigler, "The Theory of Economic Regulation" (1971) 2 Bell J. Econ & Mngm't Sci. 3.

[34] Mancur Olson, *The Logic of Collective Action* (Cambridge, MA, Harvard University Press, 1965), chapter I, esp. pp. 22–36 (hereinafter *Logic*).

[35] Such minorities are often excluded from this public choice definition of small groups. Although numerically inferior relative to the larger community, their organizational costs may be relatively higher than those of the majority. But there could be political circumstances in which they could, in fact, achieve political influence that outweighs their relative size (see Bruce A. Ackerman, "Beyond Carolene Products" [1985] 98 Harv. L. Rev. 713, suggesting that not all minorities are "discrete and insular" in the sense of *United States* v. *Carolene Prods. Co,* 304 US 144, 152–3 n. 4 [1938]).

groups is inversely related to their size. Hence, all things being equal, a smaller group of, say, producers or employers will obtain collective goods more efficiently and more quickly than larger groups of consumers or employees. This does not suggest that the small group will achieve the optimal amount of its potential collective good. Collective losses can still result from the unequal incentive of individuals within that group to contribute to the collective effort (determined by the relative stake each individual has in the collective pie). But it does suggest that "the larger the group, the farther it will fall short of providing an optimal amount of a collective good."[36] The relative strength of the smaller groups will secure their disproportionate share of the aggregate societal welfare and will externalize part of their production costs to the members of the larger groups.

Another important explanation for biased, pro small group political decisions is the relatively better opportunities of such groups to gather relevant information, which is a collective good.[37] According to this explanation, the better organizational capabilities of small groups, provided by their members' greater gains from in-group cooperation, are translated into their relative edge in obtaining and assessing information on policies. The monitoring of politicians and bureaucrats prompts them to bias policy in favor of those who can appreciate their efforts. The larger body of ill-informed voters would hardly notice their relative loss or else would attribute it to random factors. It has been shown, for example, that in developed countries, where the agricultural sector is relatively small, policies are biased in its favor, whereas in developing countries the relatively large agricultural sector is heavily taxed.[38] The information rationale – "knowledge is power" – suggests that these biased policies stem not necessarily from lobbying, but from (or also from) collective monitoring of the policies adopted.[39] This rationale also explains the growing influence of Non-Governmental Organizations (NGOs)

---

[36] Olson, *Logic*, note 34, at p. 35.

[37] Susanne Lohmann, "An Information Rationale for the Power of Special Interests" (1998) 92(4) Am. Pol. Sci. Rev. 809. See also Michael D. Rosenbaum, "Domestic Bureaucracies and the International Trade Regime: The Law and Economics of Administrative Law and Administratively-Imposed Trade Barriers" (Discussion Paper No. 250 1/99, The Center for Law, Economics and Business, Harvard Law School) (administrative procedures lowering the costs of access to information shift power over policy-making from more organized to less organized groups).

[38] See Gary S. Becker, "A Theory of Competition among Pressure Groups for Political Influence" (1983) 98 Q. J. Econ. 371 at 385.

[39] See Lohmann, "Information Rationale," note 37, at p. 812.

that advance the cause of the larger groups by promoting, for example, human rights or protection of the environment. The information they gather and disseminate improves the effectiveness of monitoring bodies such as the legislature and reduces the incentive to adopt policies that are biased against the larger groups.[40]

As Mancur Olson elaborates in *The Rise and Decline of Nations*,[41] smaller groups could organize themselves faster within the nascent Westphalian system of sovereign states and could use the state as an instrument for obtaining a disproportionate share of their resources.[42] And indeed, the political institutions of the emerging nations reflected this skewed power relationship between the smaller and larger groups. The constitution protected the smaller groups' share from the majority vote, by recognizing their holdings as protected property rights[43] and did not restrict the opportunities of the small groups to influence politicians and bureaucrats.[44]

Domestic political dynamics (and, consequently, transnational political dynamics) involve not only large and small groups vying for political influence; they include also politicians, bureaucrats, and judges whose own interests and preoccupations must also be taken into account. Politicians, whose immediate interest is election or re-election, broker public goods in exchange for campaign contributions or other political support (if they happen to function in democratic regimes) or personal financial gains (if they operate in non-democratic regimes). The bureaucracy is immune to the immediate influence of interest groups only if it is properly insulated from the political system. Political control of the appointment process is usually the most effective and pervasive way to ensure

---

[40] See Helen V. Milner, *Interests, Institutions, and Information: Domestic Politics and International Relations* (Princeton, NJ, Princeton University Press, 1997), pp. 247–8 (hereinafter – *Interests*).

[41] Mancur Olson, *The Rise and Decline of Nations* (New Haven, Yale University Press, 1982) (hereinafter – *Rise and Decline*).

[42] Olson, *Rise and Decline*, chapter 3.

[43] See the discussion on the constitution as a tool to protect property from majority decisions in *The Federalist* no. 10 (James Madison, 22 Nov., 1787); William M. Landes and Richard A. Posner, "The Independent Judiciary in an Interest-Group Perspective" (1975) 18 J. Law & Econ. 875 (presenting constitutional guarantees as securing legislative deals among diverse interest groups).

[44] Marx, of course, has made a stronger claim, namely, that these small groups, the bourgeois, invented the state system to exploit the masses. For Marxist-oriented historiography of the emergence (and possible demise) of the nation-state, which corroborates the Olsonian thesis, see Ernest Gellner, *Nations and Nationalism* (London, Cornell University Press, 1983); Eric J. Hobsbawm, *Nations and Nationalism since 1780* (2nd edn, New York, Cambridge University Press, 1992).

a submissive bureaucracy,[45] although the private market, where many regulators eventually end up, is another powerful source of influence. In many countries, bureaucracies are not insulated and, hence, tend to reflect the interests of the politicians and of interest groups. Judges are even less susceptible to such pressures, if the ideal of independence is, in fact, being observed.[46] This insulation of bureaucrats and judges from political influence provides a useful tool for the competing interest groups. In matters where shifting policies due to fluctuating political influence are undesirable, groups may agree to establish state agencies comprised of bureaucrats independent from political influence. Thus, for example, by relegating the management of the national monetary system to administrative agencies – the central banks – rival domestic groups have solved a difficult collective-action problem.[47] Similarly, constitutional guarantees against, for example, the taking of property have also been used to secure deals between domestic groups.[48]

But this understanding does not necessarily suggest that bureaucrats and judges who are not controlled by politicians necessarily pursue what they deem to be the "national interest" of their state. Their decisions may reflect certain biases. Thus, in addition to personal status and comfortable income, bureaucrats may be motivated by an interest in ensuring for themselves (and their institutions) greater discretion in the allocation of their budgets, thereby gaining latitude in implementing policies as they see fit.[49] Judges, on the other hand, may hesitate to formulate

[45] Randy Calvert, Mathew D. McCubbins, and Barry R. Weingast, "A Theory of Political Control and Agency Discretion" (1989) 33 Am. Jur. Pol. Sci. 588.

[46] See J. Mark Ramseyer, "The Puzzling (In)dependence of Courts: A Comparative Approach" (1994) 23 J. Leg. Stud. 727 (analyzing systems in which judicial dependency on the political branches results, in fact, from the politicians' interest in controlling the appointment, assignment, and promotion of judges).

[47] Geoffrey Garrett and Peter Lange, "Internationalization, Institutions and Political Change" in Robert O. Keohane and Helen V. Milner (eds.), *Internationalization and Domestic Politics* (New York, Cambridge University Press, 1996), p. 48 at pp. 66–9.

[48] See *The Federalist* No. 78, at 468 (Hamilton, 28 May, 1788) (explaining the rationale of judicial review); Landes and Posner, "Independent Judiciary," note 43 (presenting constitutional guarantees as securing legislative deals between diverse interest groups).

[49] On the objectives of bureaucrats, see William A. Niskanen, *Bureaucracy and Public Economics* (Brookfield, VT, E. Elgar, 1994), especially pp. 274–5. See also Ronald Wintrobe, "Modern Bureaucratic Theory" in Dennis C. Mueller (ed.), *Perspectives on Public Choice* (New York, Cambridge University Press, 1995), p. 429 (hereinafter – *Perspectives*). On the private benefits (mainly income, prestige, and peaceful life) and institutional benefits (larger and more discretionary budgets) that accrue to bureaucrats from participation in international organizations (which, in turn, account for their proliferation), see Bruno S. Frey, "The Public Choice of International Organization in Perspectives" in Mueller, *Perspectives* at p. 106–23.

an independent view of what "national interests" entail and, in their judgment, defer to the visions of the politicians and the bureaucrats, especially in matters involving the state's international relations. They tend to refrain from attaching any strings to the external activity of either the government or individual actors and find a myriad of ways to rebuff challenges to such activities, sometimes despite clear language to the contrary. As Roger Cotterrell observed, "judges...as state functionaries, *cannot* neglect considerations of state interests and these may, on occasion, demand that doctrinal niceties be given short shrift in order to meet particular governmental emergencies."[50]

This strong judicial deference hints at the strategy small groups have developed to shield themselves against the vagaries of democratic vote. As William Landes and Richard Posner have observed, constitutional guarantees are tools for securing the interests of smaller interest groups, notably, their property rights.[51] But no constitution is beyond legislative and judicial interpretation and immune to popular amendment to the detriment of small groups. In contrast, international law, and the judiciary's deference in international matters, offer an even better guarantee to the smaller groups' interests: the exit option.[52] And so, when larger groups have managed to mobilize and impose domestic restrictions on producers and employers, such as antitrust regulations and higher labor standards, the smaller groups have reacted by shifting their activities to foreign markets and societies.[53] When domestic groups attempted to regulate these extra-territorial activities through so-called "long arm statutes," international law provided the ultimate shelter. It enables actors to act through international organizations (IOs) that perform economic activities under the umbrella of an international agreement between states. These activities are immune to national regulation and even unencumbered by the international obligations imposed on the state parties.[54] As will be elaborated in the next chapter,

---

[50] Roger Cotterrell, *The Sociology of Law* (2nd edn, London, Butterworths, 1992), p. 235.

[51] See note 43.

[52] See Albert O. Hirschman, *Exit, Voice and Loyalty* (Cambridge, MA, Harvard University Press, 1970); cf. Richard A. Epstein, "Exit Rights under Federalism" (1992) 55 *Law and Contemporary Problems* 147 (discussing exit as a check on states power in federal systems).

[53] See John O. McGinnis, "The Decline of the Western Nation State and the Rise of the Regime of International Federalism" (1996) 18 Cardozo L. Rev. 903 at 910–11 (noting that the transformation of the Western nation-state into a welfare state accelerated its decline).

[54] See chapter 4, note 64 and accompanying text.

the laissez-faire nature of international law continues to enable smaller groups to evade national regulations and to exploit global commons. In fact, these smaller groups have had greater influence on the development of international law than on the development of domestic law, mainly because in the international arena, investments in information gathering and assessment are particularly high. Since ancient times, international negotiations have always enjoyed relative secrecy tolerated by a public and judicial vision of an ominous Machiavellian–Darwinian environment within which governments transact to ensure their very survival. Ideals of democracy and the rule of law are deemed foreign in such an anarchic and intimidating global environment. Thus, shielded from public and judicial scrutiny, small group influence on the government's conduct of external affairs has been even more significant than their domestic influence.

The systemic failures of Westphalia: efficiency, democracy, human rights, and social welfare

This look through the "veil of sovereignty" in the context of transboundary resources suggests, above all, the inaccuracy of the Westphalian paradigm. More importantly, it reveals the negative ramifications of the constitutive legal framework supported by that paradigm. This legal framework – consisting of the state-based system – is responsible for negative ramifications from the perspectives of efficiency, democracy, human rights, and social welfare.

Efficiency

As we learned from collective-action theory, to manage transboundary resources efficiently and sustainably, states must commit to pursuing long-term cooperation.[55] The fragmented, transnational view of the domestic political process and of national institutions gives an explanation for the various domestic and international collective-action failures. In contrast to the "village republics" of southern India and ancient Palestine, where each individual farmer had a voice in the collective process of resource management, contemporary national processes permit key political actors to impose severe costs on less effective actors and benefit the more influential ones. Because the more effective actors

---

[55] See discussion at chapter 2.

tend also to be those less interested in the long-term viability of the resource, they are less inclined to commit themselves (through their respective states) to long-term cooperation. Corporate actors, who can relatively cheaply relocate their activities and investments, have relatively less interest in the future of specific resources, be they domestic or transboundary. Only when the resource is unique, such as the sturgeon in the Caspian Sea, and its corporate owner has not diversified its activities, will the corporate actor adopt the long-term perspective.

These influential domestic actors share common interests with bureaucrats, who also have a rather short-term perspective when they focus on their personal conditions, the budgets they manage during their terms in office, or even their future careers in the private market. Politicians (unless organized in closely knit political parties with a long-term interest in survival) usually have an even shorter-term perspective, which extends little further than the coming elections, and they are even more accommodating to the corporate actors' interests. This influential coalition of domestic actors tends, therefore, to pursue policies that are at variance with the long-term interests of the general constituency. The outcome of such tilted power relations is an efficiency loss due to inefficient management of the resource.

Democracy

The relative edge enjoyed by small groups in capturing gains through the domestic political process poses a serious challenge to the idea of democracy. The tilted domestic power relationship is not, of course, unique to the issue of transboundary resources management. It exists also in many other domestic contexts. This general problem is, however, exacerbated in the context of transboundary resources management because such management requires coordination at the inter-governmental level. The more that cooperation is required, the less national institutions and constituencies are autonomous in making their decisions.[56] And while the general public has a lesser voice when it comes to inter-governmental negotiations and the conclusion of treaties, small interest groups enjoy even greater relative leverage. These small groups influence states' external policies through their involvement in the process of treaty negotiation and ratification. This process, described in

[56] See David Held, *Democracy and the Global Order* (Stanford, CA, Stanford University Press, 1995).

the next chapter, permits very little public scrutiny of the negotiators' acts and omissions, because the ratification process does not allow for amendments and leaves alternatives unexplored. Even the domestic debate on ratification often remains clouded, as the access the public and legislators have to information concerning the international negotiations is invariably very limited. Little is known as to which options were offered and discussed, as negotiators have little incentive to provide accurate information to the general public on their performance. Negotiators are responsible, through the treaty-ratification process, for what they have formally agreed upon; they are rarely rebuked for their missed opportunities simply because those opportunities often remain unknown to the public.[57] If we recall that alongside this deficient public scrutiny of treaties is a lack of judicial enthusiasm to scrutinize them,[58] we must come to the conclusion that democratic safeguards for ensuring the executive internalization of voter preferences regarding the state's external relations are severely handicapped. With respect to transboundary resource issues, there are few institutional oversight mechanisms that can correct this disadvantage and provide voice to the less enfranchised domestic interest groups[59] or to the interests of future generations.

Human rights

The transnational paradigm explains the tendency of governments to prefer the short-term gains of particularly strong small domestic groups at the expense of larger groups, particularly weak minority groups, as well as future generations. As described in chapter 1, disrespect for human rights and minority cultures has also been strongly associated with national management of water and other natural resources. Over the last few decades, the damming of rivers has caused the uprooting of millions of people, whose dwellings, fertile farmlands, forests, and wildlife were flooded. Resettlement schemes failed to rehabilitate the evacuated peoples, who suffered great physiological, psychological, and socio-cultural stress. Likewise, numerous individuals and communities

---

[57] See George W. Downs and David M. Rocke, *Optimal Imperfection?* (Princeton, NJ, Princeton University Press, 1995), chapter 3.

[58] See notes 50–1 and accompanying text; see also discussion at chapter 4, notes 14–33 and accompanying text.

[59] This is the logic behind John Hart Ely's theory of judicial review. See John Hart Ely, *Democracy and Distrust* (Cambridge, MA, Harvard University Press, 1980).

suffered from the drying up of floodplains and marshlands, another oft-neglected outcome of river impoundment.[60]

The entrance of foreign investors into developing markets tends to exacerbate the problem. The desire to cater to the interests of these investors and the domestic small group associated with them leads to disregard for the rights and interests of the larger population and, especially, for the rights of minority groups. Environmental tragedies throughout the developing world, from the massive gas release in Bhopal, India, through the major dumping of oil into the rivers of Ecuador and Peru, to the oil spills in Nigeria, irrevocably affect the lives of indigenous peoples. They occur because the multinational corporation that is directly responsible and the local politicians and bureaucrats who are indirectly responsible act rationally. The multinational company has a long-term interest in its investments, not in the future of a particular transboundary resource. Domestic politicians and bureaucrats pursue their own short-term goals. The courts of those states often lack the power to constrain the multinational corporation. At the same time, the court in the state where the company is domiciled is faced with the concern that a costly decision against the company would prompt it to relocate to another jurisdiction, a move that would entail the loss of jobs and revenues. These courts, both in the case of the Bhopal gas leak[61] and the Ecuador and Peru rainforest and river pollution[62] refused to entertain suits for compensation, on the basis of the doctrine

---

[60] For such effects as the displacement of inhabitants, destruction of their tribal way of life, loss of fertile lands beneath the reservoir, and loss of floodplain agriculture and fisheries downstream in the estuaries and the near-shore area, see chapter 1, notes 35–46 and accompanying text.

[61] In this case, lawsuits were brought by both individual claimants and the government of India in the US District Court of the Southern District of New York against the US parent of Union Carbide India (Union Carbide Corp.). See *In re Union Carbide Corp. Gas Plant Disaster at Bhopal, India* in December 1984, 634 F. Supp. 842 (SDNY 1986), *modified,* 809 F.2d 195 (2d Cir. 1986), *cert. den.,* 484 US 871 (1987). Despite the Indian government's assertion that Indian courts were unable to cope with the magnitude of this claim, the court opined that litigating in India would better respect Indian sovereignty and judicial self-sufficiency. *Ibid.,* at 867. The real underlying issue was, of course, the possibility of obtaining the higher US standards of damages including punitive damages.

[62] Two class action suits filed in US courts by Ecuadorian and Peruvian citizens against Texaco alleged that Texaco had severely polluted rainforests and rivers in Ecuador and Peru as a result of its oil exploitation activities in Ecuador. The suits were dismissed on grounds of *forum non conveniens* and international comity. See *Aquinda v. Texaco, Inc.,* 945 F. Supp. 625 (SDNY 1996), *vacated,* 157 F.3d 152 (2d Cir. 1998); *Sequihua v. Texaco, Inc.,* 847 F. Supp. 61 (SD Tex. 1994).

of *forum non conveniens* or "international comity." Thus, the long-term externalities of the multinational company are borne by the local general population (including future generations) and, particularly, by the underrepresented indigenous peoples who live in the affected region.

Social welfare

The Westphalian-based legal order leads to social welfare losses due to interstate competition for foreign investors and investments. The process of globalization further underlines this phenomenon. As a result of the decreasing costs of relocating investments and activities, producers, investors, and employers have fewer commitments to a specific jurisdiction. They can, therefore, exploit fully the global prisoner's dilemma game, a position in which many nations are placed, a race that constantly pressures national governments to reduce the standards for protecting consumers, employees, and the environment. This race benefits those groups for whom relocation is relatively cheap and who could re-establish themselves in less demanding states. Lower taxation on capital and caps on public spending (as a result of the openness of monetary markets) require governments to limit their budgets, and this in many cases means less public spending on social welfare.[63] State-sponsored collective-action measures that could promote better labor standards,[64] protect consumers[65] and the environment,[66] and allocate shared resources in an optimal and sustainable way[67] are, therefore, becoming more difficult to attain. Local and multinational firms exploit these failures and

---

[63] See, e.g., Geoffrey Garrett, "Capital Mobility, Trade, and the Domestic Politics of Economic Policy', in Keohane and Milner, *Internationalization and Domestic Politics* note 47, p. 79. On the impact of monetary markets on governments' budgets, see Stephan Haggard and Sylvia Maxfield, "The Political Economy of Financial Internationalization in the Developing World" in Keohane and Milner, *Internationalization and Domestic Politics*, p. 209.

[64] Both in terms of wages and social benefits, as well as in terms of security of the workplace. See Michael Trebilcock and Robert Howse, *The Regulation of International Trade* (2nd edn, London, Routledge, 1999), chapter 16; Katherine Van Wezel Stone, "Labor and the Global Economy: Four Approaches to Transnational Labor Regulation" (1995) 16 Mich. J. Int'l L. 987 at 996–7.

[65] As to the possibility of setting up global antitrust regulations, see Guzman, "International Antitrust," note 17 (suggesting a low probability for such action).

[66] Richard L. Revesz, "Rehabilitating Interstate Competition: Rethinking the 'Race-to-the-Bottom' Rationale for Federal Environmental Regulation" (1992) 67 NYU L. Rev. 1210.

[67] See chapter 2; Elinor Ostrom, *Governing the Commons* (New York, Cambridge University Press, 1990), p. 30.

externalize a substantial part of their costs both to their fellow citizens and to foreign communities.

Conclusions: The normative implications of the transnational conflict paradigm

The transnational conflict paradigm suggests that international cooperation is less than assured and the prospects for the ample provision and optimal allocation of resources are increasingly reduced. It also highlights the root of human rights deprivation in the context of resource management and allocation. The imperative, therefore, is to find ways to contain the underlying transnational conflicts of interests that are responsible for many tragedies of the global commons. This insight further suggests that domestic interests must be regulated in ways that would limit their deleterious influence on international cooperation. This imperative is supported by the goals of efficiency, democracy, human rights, and social welfare.

The scheme, then, is to examine whether states – or rather governments representing states in the international sphere – could be made to reflect more equitably and democratically the long-term interests of their respective constituencies and to what extent international law and institutions could be instrumental in such a scheme. The claim made in the following chapters is that such an endeavor is possible. In order to promote efficient, just, and sustainable allocation of transboundary resources through collective action, it is necessary to set up mechanisms that will ensure all relevant actors a voice in decision-making, monitoring, and enforcement processes, just as was the case in the ancient "village republics." To do so, the current juxtaposition of seemingly highly regulated national procedures and quite flexible and even obscure international procedures must be eradicated and replaced. The next chapter explores the "how" question: how small groups manage to gain disproportionately more than large groups. This analysis of existing institutions and norms at the national and international level sets the stage for the search for the legal and institutional remedies that would level the political playing field and increase the potential of the long-term perspective (chapters 5 and 6).

4   The transnational conflict paradigm: structural failures and responses

The systemic failures of Westphalia

This chapter explores the systemic reasons for the relative edge that small interest groups have over larger interest groups in influencing domestic decision making and international negotiations, including decisions concerning the management of transboundary resources. This first part outlines the relevant norms, procedures, and institutions – both domestic and international – that together sustain small-group influence. Based on this analysis, the second part then suggests modalities for reducing the domination of these groups and enhancing the prospects of positive cooperation among all interested groups.

The edge that small groups enjoy in relation to the management of transboundary resources stems from the fact that the adoption processes for policies relating to resources not solely under the control of a single polity are customarily less transparent than adoption policies for wholly domestic matters. The less transparent the decision-making procedures, the greater the opportunities for the executive to camouflage its concession to interest groups. The management of transboundary resources involves treaties rather than domestic prescriptions. The negotiation, implementation, and enforcement of treaties provide vast opportunities for executive slack and, hence, for interest group gains.

Treaty negotiation and ratification

As suggested by Robert Putnam, international negotiations are structured like a two-level game: a simultaneous game played by government representatives at the international level with representatives of foreign governments and at the national level with representatives of domestic interest groups.[1] But the fact that governments play the game does not necessarily imply that they actually control its outcome. In fact, the smaller domestic interest groups enjoy a large degree of control over the international treaty-making process: first, they can influence their government's position at the negotiating table;[2] second, they can influence the legislature's attitude towards the treaty during the ratification process;[3] third, due to their influence on the government, they can influence their country's attitude towards compliance with the treaty; and fourth, and for the same reason, they can influence their country's reaction to breaches by other parties to the treaty. With their influence in the negotiation and ratification processes, these small groups are positioned to stall, water-down, or block negotiations that might encroach on their interests.[4] Similar opportunities to influence treaty outcomes exist for small elites in non-democratic regimes. Such regimes also must secure informal ratification from the elites from whom they draw support.[5]

[1] Robert D. Putnam, "Diplomacy and Domestic Politics: The Logic of Two-Level Games" (1988) 42 Int'l Org. 427 at 436. On this two-level game, see also Helen V. Milner, *Interests, Institutions, and Information: Domestic Politics and International Relations* (Princeton, NJ, Princeton University Press, 1997), pp. 247–8; George W. Downs and David M. Rocke, *Optimal Imperfection?* (Princeton, NJ, Princeton University Press, 1995); Peter B. Evans, Harold K. Jacobson and Robert D. Putnam (eds.), *Double-Edged Diplomacy* (Berkeley, University of California Press, 1993) (an analysis of eleven cases of two-level bargaining). For an argument for a "three-level game" (including an additional "transnational/ transgovernmental bargaining" as "level I") see Thomas Risse-Kappen, "Structures of Governance and Transnational Relations: What We Have Learned?" in Thomas Risse-Kappen (ed.), *Bringing Transnational Relations Back In* (Cambridge, Cambridge University Press, 1995), p. 80 at p. 300.

[2] Although it is theoretically possible that negotiators would misinform their small domestic interest groups about the nature and constraints of the international negotiations, this is highly unlikely to occur, because interested groups would invest in closely monitoring the negotiators or even demand to take part in the negotiations.

[3] The form of ratification has bearing on the position each government adopts at the international level. Thus, the differences in the ratification procedures in the United States and Britain and in the legal status of treaties within the respective domestic legal systems can explain the different attitudes in these countries toward the adoption of international environmental standards. Compare Kal Raustiala, "The Domestic Politics of Global Biodiversity Protection in the United Kingdom and the United States" in Miranda A. Schreurs and Elizabeth Economy (eds.), *The Internationalization of Environmental Protection* (Cambridge, Cambridge University Press, 1997), p. 42 at pp. 48–52 (comparing the different attitudes of the two countries in negotiating international biodiversity standards).

[4] See Milner, *Interests*, note 1, chapter 3.

[5] See Peter B. Evans, "Building an Integrative Approach to International and Domestic Politics in Double-Edged Diplomacy" in Evans, *Doubled-Edged Diplomacy*, note 1, at p. 397, pp. 415–16; Kurt T. Gaubatz, "Democratic States and Commitment in International Relations" (1996) 50 Int'l Org. 108.

It is important to understand why the processes of treaty negotiation and ratification are even more vulnerable to interest-group influence than regular legislation is. In contrast to the relatively transparent and accessible regular legislation process, the performance of international negotiators is less open to serious domestic scrutiny and effective democratic deliberation. Often, the government includes representatives of interest groups on the negotiating team. This enhances the representation of these groups in the negotiations, although it also co-opts the groups and can reduce their ability to resist the treaty's ratification.

Unlike legislative proposals, which the legislature controls from the initial steps of the introduction of bills, to the discussion of amendments, and through to the final product, the treaty to be ratified is a completed transaction whose alternatives cannot be explored fully and which is not subject to alterations.[6] In those states where legislative approval is required for treaty obligations to take effect, the post-hoc ratification process allows the legislature far less scrutiny and control compared to the regular legislation process.[7] The sequential process of treaty ratification allows the government a relatively free hand in setting the agenda, formulating the policies, and choosing among the alternatives.[8] The *ex post* "take it or leave it" option presented to the ratifying body presents a hurdle often too high for opponents of the treaty or those who seek to amend it.

The ratification process allows less public scrutiny not only because of its sequential nature. Transparency is further restricted because the treaty ratification procedure allows for less public scrutiny of the norm being prescribed than does the enactment procedure for regular legislation. Under the ratification process in many of the countries following the parliamentary system of government – including the United Kingdom, some of the Commonwealth countries, and Scandinavian

---

[6] For an analysis of the political advantage of presidents over the legislature in holding exclusive power to initiate, to make take-it-or-leave-it proposals, and to control information, see Matthew Soberg Shugart and John M. Carey, *Presidents and Assemblies* (Cambridge, Cambridge University Press, 1992), pp. 139–40.

[7] See Stefan A. Riesenfeld and Frederick M. Abbott, "Foreword: Symposium on Parliamentary Participation in the Making and Operation of Treaties" (1991) 67 Chi. Kent L. Rev. 293 at 303; for a comparative survey of the practices of treaty ratification, see Francis G. Jacobs and Shelley Roberts (eds.), *The Effect of Treaties in Domestic Law* (London, Sweet & Maxwell, 1987); and the symposium published in (1991) 67 Chi. Kent L. Rev. 293–704.

[8] As Kenneth Arrow has demonstrated, the agenda-setter can determine the outcome of the vote when there are more than two options: Kenneth J. Arrow, *Social Choice and Individual Values* (New York, Wiley, 1951).

countries – it is the government, and not the legislature, that ratifies treaties (whereas their incorporation into domestic law is effected by a statute or an administrative regulation).[9] The fact that the government enjoys a majority in Parliament or that high political costs are entailed in bringing the government down usually thwarts in-depth public deliberation of the treaty in Parliament.

In the United States, the interplay between Congress and the Executive in treaty ratification is more complicated than in parliamentary systems. On the one hand, the constitutional requirement of a two-thirds majority in the Senate for "advice and consent" to treaties places veto power in the hands of a minority of thirty-four senators,[10] thereby enabling small groups to direct their efforts at only a small number of legislators. On the other hand, such a strategy on the part of small groups is compromised by the widely accepted practice of presidential bypass of the Senate by concluding international obligations (domestically referred to as "executive agreements") with only Congress' approval or unilaterally by the President.[11] "Presidents have asserted a broad authority to make many other international agreements [in addition to recognition of states and armistice agreements], at least in the absence of inconsistent legislation or of Congressional action restricting such agreements."[12] Small groups, therefore, need to invest in influencing the Executive. In recent years, a so-called "fast-track" procedure has developed with respect to trade agreements in which the President agrees to involve Congress in the negotiations phase in return for a bicameral

---

[9] For the British procedure, see Rosalyn Higgins, "United Kingdom" in Jacobs and Roberts, *Effect of Treaties*, note 7, at pp. 123–4; Lord Templeman, "Treaty-Making and the British Constitution" (1991) 67 Chi. Kent L. Rev. 459; for the situation in Denmark, see Claus Gulmann, "Denmark" in Jacobs and Roberts, *Effects of Treaties*, p. 29, at pp. 29–30; on Iceland, see Ragnar Adalsteinsson, "The Current Situation of Human Rights in Iceland" (1994) 61/62 Nordic J. Int'l L. 167 at 168–70; on Sweden, see Michael Bogdan, "Application of Public International Law by Swedish Courts" (1994) 63 Nordic J. Int'l L. 3 at 8–11.

[10] See Stefan A. Riesenfeld and Frederick M. Abbott, "The Scope of US Senate Control over the Conclusion and Operation of Treaties" (1991) 67 Chi. Kent L. Rev. 571 at 601.

[11] See *United States* v. *Belmont*, 301 US 324, 57 S. Ct. 758. 81 L.Ed. 1134 (1937); *United States* v. *Pink*, 315 US 203, 62 S. Ct. 552, 86 L.Ed. 796 (1942); *Dames & Moore* v. *Reagan*, 453 US 654, 101 S. Ct. 2972, 69 L.Ed. 918 (1981).

[12] Restatement (Third) on the Foreign Relations Law of the United States (1987), sec. 303 and cmt. g. On executive agreements, see Louis Henkin, *Foreign Affairs and the United States Constitution* (2nd edn, Oxford, Oxford University Press, 1996), pp. 215–30; Joel R. Paul, "The Geopolitical Constitution: Executive Expediency and Executive Agreements" (1998) 86 Calif. L. Rev. 671; Riesenfeld and Abbott, "Scope of US Senate," note 10, at 635–41; John H. Jackson, "United States," in Jacobs and Roberts, *Effects of Treaties*, note 7, at p. 141, pp. 142–4.

congressional commitment to vote the agreement up or down with-out amendment.[13] Congress' involvement at the negotiations stage lim-its the discretion of government negotiators during the international bargaining process and provides more voice to groups that are less in-fluential with the Executive, although the President continues to con-trol the agenda. The US example is atypical in this regard. It reflects the complex interaction between Congress and the President and the greater influence of certain interest groups on Congress than on the President.

The US courts have upheld this lack of significant legislative supervi-sion of international negotiations. In the celebrated case of *United States v. Curtiss-Wright Export Corp.*,[14] the Supreme Court affirmed that

in this vast external realm, with its important, complicated, delicate and mani-fold problems, the President alone has the power to speak or listen as a repre-sentative of the nation. He makes treaties with the advice and consent of the Senate; but he alone negotiates. Into the field of negotiation the Senate cannot intrude; and Congress itself is powerless to invade it.[15]

The court used Congress' relative lack of information as an argument against the latter's involvement in decision making:

[The President], not Congress, has the better opportunity of knowing the condi-tions which prevail in foreign countries, and especially is this true in time of war. He has his confidential sources of information. He has his agents in the form of diplomatic, consular and other officials. Secrecy in respect of informa-tion gathered by them may be highly necessary, and the premature disclosure of it productive of harmful results.[16]

The courts still adhere to this argument, although it has become more and more difficult to draw a sharp distinction between domestic and international affairs, and international negotiations tend to impose sig-nificant burdens on domestic policies and, hence affect citizens' rights and interests.

The same rationale (i.e., that the executive branch is the sole repre-sentative in this context) influences the courts in most democratic coun-tries. They have shown reluctance to assert their judicial authority to

---

[13] See Harold H. Koh, "The Fast Track and United States Trade Policy" (1992) 18 Brook. J. Int'l L. 143; Detlev F. Vagts, "Comment: The Exclusive Treaty Power Revisited" (1995) 89 AJIL 40; Riesenfeld and Abbott, "Scope of US Senate," note 10, at 637–8. See generally George A. Bermann, "Constitutional Implications of US Participation in Regional Integration" (1998) 46 *American Journal of Comparative Law* 463.

[14] 299 US 304, 57 S. Ct. 216, 1936 US Lexis 968, 81 L. Ed. 255 (1936).

[15] 299 US at 319.   [16] *Ibid.*, at 320.

review ratified treaties. As a result, treaties enjoy greater immunity from judicial review than statutes. In some countries, the constitution endows treaties with a higher status than statutes and even higher than the constitution itself, thus giving treaties immunity from judicial scrutiny.[17] In countries where treaty ratification is deemed a governmental prerogative, treaties are insulated from judicial review, somewhat paradoxically, because they have no direct effect in the domestic legal system and, hence, do not affect individual rights.[18] In states where ratified treaties share the same status as statutes, national courts are, in theory, competent to review the constitutionality of treaties. The Constitutional Courts of Italy, Germany, Japan, and the United States have asserted this authority, but at the same time emphasized that only in rare and exceptional circumstances will they exercise their power of review. The Japanese Supreme Court refused to review the Japan–US Security Treaty, noting its political nature, and declared that it would intervene only when the unconstitutionality or invalidity of the treaty is "obvious."[19] The US Supreme Court indicated its inclination to relax constitutional norms in deference to "the never-ending tension between the President exercising the executive authority in a world that presents each day some new challenge with which he must deal and the Constitution."[20] The German Constitutional Court "will spare no effort and, in fact, will go out of its way, to reconcile Germany's treaty obligations with its internal legal order."[21] The Italian Constitutional Court "took the attitude of under-valuing conflicts between treaties and the constitution."[22] In "dualistic"

[17] See, for example, article 55 of the French Constitution, article 120 of the Dutch Constitution (see "France" and "The Netherlands" in Jacobs and Roberts, *Effects of Treaties,* note 7, at pp. 42, 111 respectively).

[18] This applies to Britain (Templeman, "Treaty-Making," note 9, at 461) and other Commonwealth countries, as well as to Scandinavian countries (see references in note 7).

[19] The Sunagawa case, reprinted in 3 Int. L.R. (1960) 43. See Yuji Iwasawa, *International Law, Human Rights, and Japanese Law* (Oxford, Oxford University Press, 1998), p. 10 (arguing that "in reality, however, treaties would be unlikely to be subject to judicial review.")

[20] *Dames & Moore* v. *Reagan,* note 11, at 662. In the US, the single case in which a treaty (domestically classified as an executive agreement) was struck down as incompatible with the Constitution remains *Reid* v. *Covert,* 354 US 1 (1957).

[21] Jochen Abr. Frowein and Michael J. Hahn, "The Participation of Parliament in the Treaty Process in the Federal Republic of Germany" (1991) 67 Chi. Kent L. Rev. 361 at 385. This court has found a treaty incompatible with the Constitution only once.

[22] Giorgio Gaja, "Italy" in Jacobs and Roberts, *Effect of Treaties,* note 7, at pp. 87, 101. The Constitutional Courts of Germany and Italy also have upheld their competence to review parliamentary decisions to transfer sovereign powers to the European Union (but have never exercised it); see Paul p. Craig and Grainne de Burca, *EU Law*

states, where treaties are incorporated into domestic law by legislation and the legislation is subject to judicial review, the incorporated law can, in principle, be reviewed under the domestic constitution.[23] But after the treaty has been ratified, the courts will be reluctant to create conditions that will require the executive to breach international obligations.

Such judicial caution fits well with the general attitude of national courts to defer to the discretion of the executive in conducting the country's foreign affairs.[24] As Louis Henkin observes regarding the attitude of the US Supreme Court to matters involving foreign affairs, "Judicial review rarely asserts or spends itself."[25] More generally, Henkin suggests that "where 'balancing' an individual right against the public interest is deemed to be the constitutional order, courts treat foreign affairs differently: private rights are depreciated, while competing public needs are accorded compelling weight."[26] Judicial interference with treaty obligations is deemed by courts an uncalled-for intervention in international affairs, regardless of the domestic implications. The basic attitude of courts throughout the democratic world has been that in international affairs, "our State cannot speak with two voices on such a matter, the judiciary saying one thing, the executive another,"[27] and the executive's voice must be preferred because of the inherent "advantage

(2nd edn, New York, Oxford University Press, 1998), pp. 268–79. On the German Constitutional Court's decision with respect to the constitutionality of the Maastricht Treaty, see notes 120–2 and accompanying text.

[23] See *Horta* v. *Commonwealth of Australia* (1994) 123 ALR 1; 104 Int. L.R. 450 (1997) (examining whether legislation giving effect to the Timor Gap Treaty with Indonesia is valid under Australia's constitution).

[24] See Eyal Benvenisti, "Judicial Misgivings Regarding the Application of International Norms: An Analysis of Attitudes of National Courts" (1993) 4 Eur. J. Int'l L. 159. This comparative survey demonstrates that the phenomenon is common to all national courts and, therefore, cannot be explained by parochial theories. See also Benedetto Conforti, *International Law and the Role of Domestic Legal Systems* (Boston, M. Nijhoff, 1993), pp. 13–47. Compare Bruce Ackerman and David Golove, "Is NAFTA Constitutional?" (1995) Harv. L. Rev. 799 (explaining judicial acceptance of the constitutionality of "executive agreements" as reflecting a "constitutional moment" that occurred during the 1944 elections); Paul, "The Geopolitical," note 12 (explaining that policy and other pro-presidential policies as influenced by the Cold War rhetoric used by the Administration).

[25] Henkin, *Foreign Affairs*, note 12, at p. 134.

[26] Louis Henkin, *Constitutionalism, Democracy, and Foreign Affairs* (New York, Columbia University Press, 1990); Paul, "The Geopolitical," note 12, at 758, 763 ("the court hinted that treaties were not subject to the same constitutional limitations as acts of Congress"; at 758).

[27] *The Arantzazu Mendi*, [1939] A. C. 256, 264.

of the diplomatic approach to the resolution of difficulties between two sovereign nations, as opposed to the unilateral action by the courts of one nation."[28] Hence, only the executive's voice will be heard; not only will courts tend to abstain from reviewing international treaties (or incorporating statutes) for compatibility with domestic prescriptions, but in interpreting them, courts in general will tend to defer to the interpretation provided by the executive.[29] Furthermore, a variety of judicially developed "avoidance doctrines" permit the courts to dodge petitions to review treaties for alleged violations of domestic norms or domestic policies for alleged infringement of international norms.[30] An independent domestic judicial interpretation of international norms is most likely to occur when the relevant international norm addresses matters of solely domestic consequence with no ramifications for international relations. Thus, for example, national courts of many countries have been very creative with respect to developing indirect avenues for adopting the standards set by international human rights conventions.[31] But it is one thing to find the government obliged to respect political freedoms of fringe parties and quite another to find it in breach of a trade agreement. Thus, the same national courts that engage in productive "transjudicial communications"[32] when human rights are at stake will still follow partisan policies when external relations are at issue.[33]

This judicial deference to the executive is deeply troubling, as it permits a sizable amount of executive activity, with major ramifications for

---

[28] *United States* v. *Alvarez-Machain*, 504 US 655, 112 S. Ct. 2188, 1992 US Lexis 3679, 119 L. Ed. 2d 441 n. 16 (1992).

[29] See Benvenisti, "Judicial Misgivings," note 24, at 166–8; Conforti, *Domestic Legal Systems*, note 24, at pp. 17–20.

[30] Benvenisti, "Judicial Misgivings," note 24, at 169–73.

[31] Several scholars report a judicial inclination to adhere to international standards in the context of domestic human rights litigation. See, e.g., Harold Hongju Koh, "The 'Haiti Paradigm' in United States Human Rights Policy" (1994) 103 Yale L.J. 2391; Harold Hongju Koh, "Transnational Public Law Litigation" (1991) 100 Yale L.J. 2347; Anne Bayefsky and Joan Fitzpatrick, "International Human Rights Law in United States Courts: A Comparative Perspective" (1992) 14 Mich. J. Int'l L. 1; Eyal Benvenisti, "The Influence of International Human Rights in Israel: Present and Future" (1995) 28 Israel L. Rev. 136. In Japan adherence to international human rights standards was achieved indirectly through litigation: the failure of courts to implement those standards mobilized public opinion to seek a political solution: Iwasawa, *Japanese Law*, note 19, at pp. 308–9.

[32] Anne-Marie Slaughter, "A Typology of Transjudicial Communications" (1994) 29 U. Rich. L. Rev. 99.

[33] Meinhard Hilf, "New Frontiers in International Trade: The Role of National Courts in International Trade Relations" (1997) 18 Mich. J. Int'l L. 321 at 338–43.

domestic interests, to remain completely beyond judicial reach and effective public scrutiny. As we shall see in the next two sections, the same stance persists beyond the ratification stage, when attempts are made to unilaterally terminate a treaty or to temporarily renege on it. This stance ensures the executive branch unfettered discretion in deciding whether to comply with treaty obligations.

Ensuring treaty durability

The stronger domestic effect of treaties in comparison to statutes is enhanced by the norms that insulate treaty obligations from subsequent domestic challenges by larger political groups. Once a treaty has been ratified, there is neither opportunity to revoke it through judicial review procedures nor the possibility to terminate it through unilateral state action by, for example, a statute. Unless the ratifying state chooses to violate the treaty and suffer whatever consequences this may entail, it remains bound by the treaty until the other parties agree to modify or terminate it. Herein lies the ultimate benefit of treaties for the domestic smaller groups: treaties protect their gains even more than constitutional guarantees do. Unlike constitutional guarantees, which are vulnerable to modification by constitutional amendment, no subsequent domestic majority can render unilateral changes to the state's international obligations.[34] Future governments will, therefore, continue to be bound by the same obligations.

This immunity endowed to international treaties is grounded in a combination of constitutional and international doctrines. On the international plane, the durability of treaties is a function of the doctrine that renders irrelevant all subsequent domestic political developments as factors that can affect a state's international obligations. The strength of this doctrine was demonstrated in the September 1997 judgment handed down by the International Court of Justice ("ICJ") in the dispute between Hungary and Slovakia over the use of the Danube River.[35] In its judgment, the ICJ determined that the damming of the Danube, a mammoth project conceived in the bygone communist era, should go

---

[34] Compare with the insights of William M. Landes and Richard A. Posner, "The Independent Judiciary in an Interest-Group Perspective" (1975) 18 J. Law & Econ. 875 (discussing constitutional guarantees as securing legislative deals among politicians and diverse interest groups).

[35] Gabcikovo–Nagymaros Project (Hungary/Slovakia), Judgment, ICJ Reports 1997, p. 7, reprinted in http://www.icj-cij.org/idocket/ihs/ihsjudgement/ihsjudframe1.htm; (1998) 37 ILM 167.

ahead as planned, despite the momentous political transformations that had taken place in the two countries and the intensive and widespread popular opposition in Hungary to the project. Neither domestic political changes nor strong popular opposition could provide the basis for unilateral termination of a 1977 treaty between the two states.[36] The communist legacy, however inefficient or environmentally dangerous it may be, survived the transformation of both regimes and unilateral contradictory moves of the two governments.[37] In reaching this conclusion, the court deliberately emphasized international undertakings at the expense of domestic pressures. It rejected Hungary's claim that a "state of ecological necessity," even if such existed, precluded the wrongfulness of its unilateral suspension of the project, because Hungary could resort to negotiations to reduce the environmental risks. It similarly rejected Hungary's claims to impossibility of performance, fundamental change of circumstances, and lawful response to Czechoslovakia's earlier material breach (namely, Slovakia's construction of the provisional diversion project).[38] At the same time, the ICJ was critical also of Slovakia's moves. It found the Slovak diversion of the Danube waters a breach of its obligation towards Hungary to respect the latter's right to an equitable and reasonable share of the river.[39] Finding the agreement flexible and, therefore, renegotiable, the ICJ imposed the 1977 treaty on both parties, instructing them to negotiate its implementation and modification.[40] This judgment clearly sought to insulate international agreements from the influence of domestic politics. Even when one side breaches its obligation to renegotiate in good faith, the government of the other side cannot bow to domestic public pressure and take unilateral measures. Instead, the latter must exhaust all possible means, including through third parties, to persuade its counterpart to return to the negotiating table.

---

[36] On the development of the internal environmental–political opposition in Hungary to the planned project on the Danube River, see Fred Pearce, *Green Warriors* (London, Bodley Head, 1991), pp. 107–16 ; Judit Galambos, "Political Aspects of an Environmental Conflict: The Case of the Gabcikovo–Nagymaros Dam System", in Jurki Kakonen (ed.), *Perspectives on Environmental Conflict and International Politics* (London, Pinter Publishers, 1992), p. 72.

[37] See note 35, paras. 144–7.

[38] *Gabcikovo-Nagymaros Project (Hungary/Slavakia)* Judgment, ICJ Reports 1997, paras. 101–12.

[39] *Ibid.*, para. 78.

[40] *Ibid.*, paras 138–40. Note that the court is careful to maintain equality between the parties. It did not *require* Hungary to fulfill its obligations under the 1977 treaty, such as building a second dam, unless the parties were to agree otherwise, but, rather, required them to renegotiate the outstanding obligations in good faith.

Interestingly, this is the cumulative message of the ICJ decision, which, somewhat paradoxically, was shared by only a minority of six judges who concurred that all the unilateral measures were unlawful. The other nine judges approved the unilateral action of either one of the parties.[41] This message is captured in President Schwebel's declaration. Although he was "not persuaded that Hungary's position as the Party initially at breach deprived it of a right to terminate the Treaty in response to Czechoslovakia's material breach," the President joined the majority in imposing the resuscitated agreement on the parties.[42]

The international legal doctrine is reinforced in many countries by either a constitutional guarantee or a doctrine that insulates treaties from attempts by domestic groups to force the government to breach treaty obligations. Theoretically, except in the few states where treaty obligations enjoy superior normative status,[43] statutes can circumvent treaty obligations. The legislature can pass a conflicting law and thereby expose the state only to external sanctions. But a judicially developed doctrine could preempt most such efforts. There is a rule of statutory interpretation, accepted in many jurisdictions, that provides that statutes should be interpreted, to as great an extent as possible, in conformity with the state's international obligations.[44] Although the declared underlying rationale of this rule is the presumed intention of the legislature, courts have resorted to this rule in spite of a rather clear indication from the legislature that it wants the treaty breached. A case in point is the "sad case of the PLO mission."[45] The American federal judge in this case disregarded the clear intention of Congress to breach the Headquarters

---

[41] Five judges approved Czechoslovakia's implementation of the provisional solution, whereas the four other judges approved Hungary's termination of the treaty (*Ibid.*, para. 155).

[42] *Ibid.* (declaration of President Schwebel).   [43] See notes 19–22.

[44] For a comparative survey of this presumption, see M. Hayward, "International Law and the Interpretation of the Canadian Charter of Rights and Freedoms: Uses and Justifications" (1985) 23 U. West. Ontario L. Rev. 9 at 13–16 (Canada); see Claus Gulmann, "Denmark," in *Effects of Treaties,* note 7, at p. 36 (Denmark); *R. v. Home Secretary, Ex p. Brind* [1991] 1 A. C. 696, 760 (H. L.), *Derbyshire County Council* v. *Times Newspapers Ltd* [1992] 3 WLR 28, 44 (C. A.) (United Kingdom); Jochen Abr. Frowein, "Germany," *Effects of Treaties,* note 7, at pp. 63, 68–9 (Germany); *Kubik Darusz* v. *Union of India* et al., 92 ILR 540 (1993) (SC) (India); *Custodian of Absentee Property* v. *Samra* et al., 10 PD. 1825, 1831, 22 ILR 5 (1956) (Israel); Giorgio Gaja, "Italy," *Effects of Treaties*, note 7, at pp. 100–1 (Italy); *Minister of Defence, Namibia* v. *Mwandinghi*, 91 ILR 341 (1993) (HC) (Namibia); *State* v. *Ncube* et al., 91 ILR, 580 (1993) (SC) (Zimbabwe).

[45] *US* v. *The Palestine Liberation Organization* et al., 695 F. Supp. 1456 (1988); see W. Michael Reisman, "An International Farce: The Sad Case of the PLO Mission" (1989) 14 Yale J. Int'l L. 412 at 429–32.

Agreement between the US and the UN and to prevent the PLO leader from arriving in New York. The court found the law vague enough to permit an interpretation that would undermine its aim and prevent an international conflict.

The outcome of the combination of the international and constitutional doctrines is to insulate governments from domestic challenges that force them to renege on treaty obligations against their will. The smaller groups whose interests are secured by a given treaty can be certain that their gains will last until the larger domestic groups within all the relevant states join forces in a coordinated effort to modify the treaty obligations. They also know, as we shall see in the next section, that when an international breach *is* within their interests, their governments will not be effectively constrained from making it. In other words, the norms pave a one-way street: governments can hardly be *forced* to renege on treaty obligations, but when they choose to do so, they cannot be *prevented* from doing so. This is exactly what small groups who invest in controlling their governments want to ensure. As discussed in chapter 3, the more leeway a government enjoys, the more opportunities there are for small groups to steer policies in their favor.

Providing escape clauses for unilateral defections

All international treaties with respect to transboundary resource management, and, in particular, water management, are "relational" in the sense that they create relations between the parties that extend well into the future.[46] During the lifetime of such treaties, conditions often change, and therefore, state negotiators take pains to ensure that treaty obligations will provide efficient mechanisms for adjustment to such changes. States prefer to retain control over their reactions to such changes, instead of conferring authority on international institutions to determine what adjustments are necessary. State discretion is secured through ambiguous texts, insufficient monitoring tools, or sub-optimal enforcement mechanisms. This discretion is sought by the domestic

---

[46] On the characteristics of "relational contracts," see Ian R. Macneil, *The New Social Contract: An Inquiry into Modern Contractual Relations* (New Haven, Yale University Press, 1980); Ian R. Macneil, "The Many Futures of Contract" (1974) 47 S. Cal. L. Rev. 691; Ian R. Macneil, "Economic Analysis of Contractual Relations: Its Shortfalls and the Need for a 'Rich Classificatory Apparatus'" (1981) 75 Nw U.L. Rev. 1018; Charles J. Goetz and Robert E. Scott, "Principles of Relational Contracts" (1981) 67 Va. L. Rev. 1089; Alan Schwartz, "Relational Contracts in the Courts: An Analysis of Incomplete Agreements and Judicial Strategies" (1992) 21 J. Leg. Stud. 271.

groups within the state parties whose interests may be affected by future compliance with treaty obligations. Thus, as George Downs and David Rocke explained, international trade law has weak enforcement norms to accommodate the uncertain future demands of domestic interest groups.[47] Downs and Rocke extended their observation to other types of international agreements, where states want to be able to respond periodically to domestic interests.[48] Enforcement norms are designed to encourage the parties to observe the agreement most of the time and thereby prevent, for example, trade wars, "but low enough to allow politicians to break the agreement when interest group benefits are great."[49] Retaliation by co-signatories for a breach will usually not be targeted only against the interest group responsible for the breach, and therefore, the costs of the breach will be externalized to other domestic groups as well. This is demonstrated by the trade disputes between the United States and the European Union over EU restrictions on the importation of genetically modified food, hormone-treated beef, and bananas from Central America, and the retaliatory measures adopted by the US, targeting specific commodities imported from the EU.[50] Louis Henkin's famous observation that "almost all nations observe almost all principles of international law and almost all of their obligations almost all of the time"[51] holds an important insight with regard to the vitality of international law; yet this vitality hinges on the escape routes international law provides for all governments sometimes.

Note that escape clauses that can offer impunity for a defecting government are less likely to be of advantage to a legislature seeking to renege on a treaty. Such escape clauses are more likely to reflect the interests of small groups. The interests of larger groups, such as in sound environmental policies or labor conditions, are less likely to permit defection with impunity. The legislature is also less capable than the government of monitoring other states' performance and identifying those states' violations, a key tool – if only rhetorical – for justifying one's

[47] Downs and Rocke, *Optimal Imperfection?*, note 1, at p. 88.
[48] *Ibid.*, at pp. 88–104.   [49] *Ibid.*, at p. 77.
[50] The WTO Appellate Body found the EU in breach of trade agreements in both the banana and beef hormone disputes. See the WTO Appellate Body Reports, *European Communities – Regime for the Importation, Sale and Distribution of Bananas*, Rep. No. WT/DS27/AB/R, 9 September 1997; *EC Measures concerning Meat and Meat Products (Hormone)*, Rep. No. WT/DS26/AB/R, WT/DS48/AB/R, 16 January 1998).
[51] Louis Henkin, *How Nations Behave* (2nd edn, New York, Columbia University Press, 1979).

own breach to the international community and international tribunals. Moreover, because breaches are immediately followed by international negotiations and, possibly, adjudication, the government is more capable than the legislature of finessing the consequences of a breach (or externalizing the costs to the larger public). Due to this relative edge of the government over the legislature, it is more likely that the government will initiate unilateral defection from international obligations, whereas the legislature will remain passive.

The small domestic interest groups that invest in monitoring the government and influencing its decision making do not seek protection from the courts against their government's decision to renege on treaty obligations against their wishes. They, as opposed to groups who need the court's protection,[52] fare better by relying on the executive. Their influence on the government, which is bolstered by the international norms on treaties that serve their interest in flexibility,[53] provides sufficient defense against a governmental decision to renege. In fact, these small groups are interested in ensuring that domestic courts remain uninvolved and refrain from demanding that a reneging government (that had decided to renege at the demand of the small groups) comply with treaty obligations. And indeed, the courts conform to this expectation. Petitions for injunctions against the executive's breach of international obligations have often been brought before national courts. Such suits have claimed that domestic policies – whether in the form of statutes or administrative decisions – are incompatible with general international law[54] or with specific treaty obligations.[55] Other suits have petitioned against the domestic recognition of acts of foreign governments, such as expropriation of private property abroad or the alleged

---

[52] The protection of interest-group gains against subsequent legislation is a well-known rationale for judicial independence. See Landes and Posner, "Independent Judiciary," note 34; J. Mark Ramseyer, "The Puzzling (In)dependence of Courts: A Comparative Approach" (1994) 23 J. Leg. Stud. 721.

[53] See notes 46–51 and accompanying text.

[54] This is notably the case with efforts to invoke international human rights standards in domestic courts. See *Sale* v. *Haitian Centers Council,* 509 US 155; 113 S. Ct. 2549; 1993 US Lexis 4247; 125 L. Ed. 2d 128 (1993); and references in notes 23, 31, 32.

[55] See, e.g., *US* v. *Alvarez-Machain,* note 28 (incompatibility of a forced abduction of a Mexican citizen with the US–Mexican extradition treaty). For a similar decision by a chamber of the German Federal Constitutional Court, see the decision of 17 July 1985 (EuGRZ 1986, 18, at 20) (*Stocke* v. *the Federal Republic of Germany*). See Note, "Judicial Enforcement of International Law against the Federal and State Governments" (1991) 104 Harv. L. Rev. 1269.

annexation of territory, which international law regards as illegal.[56] As previously suggested,[57] the prevailing attitude of national courts to such claims has been to deny such petitions, as judges invariably choose to defer to the executive, unless the matter is unrelated to the state's foreign relations (as in cases involving domestic enforcement of international human rights law). Courts in virtually all democracies have shown great ingenuity in creating an arsenal of "avoidance doctrines" that enable them to align their judgments with national interests. They avoid questioning the lawfulness of the executive's activities on the international plane or the fruits of its negotiations.[58] In some jurisdictions, notably the US, this jurisprudence provides that the executive may lawfully violate international customary law.[59] It is also widely accepted in this jurisprudence that the executive may terminate treaties unilaterally and without parliamentary or judicial review, unless, of course, a treaty has been incorporated into domestic law by statute.[60] When human rights are involved, a few decisions have suggested that such termination will be subject to judicial review for procedural fairness if the termination infringes individual expectations.[61]

---

[56] A notable example of the first type of suit is the case of *Banco Nacional de Cuba* v. *Sabbatino*, 376 US 398 (1964), and its progeny (lawfulness of expropriation of private assets of US citizens in Cuba); *Horta* v. *Commonwealth*, note 23, is an example of the second type (the Australian High Court rejected a challenge to the validity of the Timor Gap Treaty between Australia and Indonesia, the illegal occupant of East Timor from 1975 to 1999).

[57] See notes 14–33.   [58] *Ibid.*

[59] *Garcia Mir* v. *Meese*, 788 F.2d 1446 (11th Cir, 1986) (cabinet officers, in addition to the President, may violate international customary law); Louis Henkin, "International Law as Law in the United States" (1984) 82 Mich. L. Rev. 1555 at 1568; Agora: "May the President Violate Customary International Law?" (1986) 80 AJIL 913–47.

[60] On the US law, see *Restatement*, note 12, sec. 339(b) ("Under the law of the United States, the President has the power ... to make the determination that would justify the United States in terminating or suspending an agreement because of its violation by another party or because of supervening events, and to proceed to terminate or suspend the agreement on behalf of the United States"); on German law, see Frowein and Hahn, "Participation of Parliament," note 21, at p. 363 (treaty termination power is considered to be within the exclusive domain of the executive). In those countries that do not require parliamentary approval of treaties, executive termination power is obvious.

[61] *Higgs* v. *Minister of National Security*, 144 SJLB 34 (Privy Council, 14 December 1999) (per Lord Hoffmann): "The executive cannot depart from the expected course of conduct unless it has given notice that intends [sic] to do so and has given the person affected an opportunity to make representations"; see also *Thomas* v. *Batiste*, [1999] 3 WLR 249 (Privy Council), Minister for Immigration & Ethnic Affairs v. Teoh, (1995) 128 ALR 353, 104 ILR 460 (1997) (Australian High Court).

Some have suggested that a probable explanation for this prevalent judicial attitude is the lack of democratic legitimacy underlying
international treaties and other norms.[62] Because of the absence of
domestic democratic control over the formulation of international
treaties and custom, goes the argument, international obligations are
interpreted as not constituting an integral part of the domestic law and,
hence, outside the proper scope of judicial scrutiny. But this explanation
is inadequate because it merely reinforces what we already know and
its conclusion further exacerbates the problematic democratic deficit.
It only highlights the undemocratic consequences of the courts' hesitation to review these matters: due to this judicial deference, international
treaties, which often have significant, if indirect, domestic ramifications,
bypass the courts' muster simply because they have bypassed the regular
legislative procedures.

Moreover, the courts' deference to the executive is not motivated by a
concern over the absence of democratic control. Rather, it is compelled
by the same global inter-jurisdictional competition that affects the other
branches of government. An assertive court that is ready and willing to
enforce international norms on a recalcitrant government (which, for example, refrains from imposing environmental standards on polluting industries) reinforces the sanctions for dodging international obligations
and thereby increases their cost.[63] Potentially affected firms may, therefore, decide to relocate to another jurisdiction, where courts are more
lenient. For this same reason, national courts have no incentive to encroach on the laissez-faire philosophy of international law or to develop
stringent standards for the activities of locally based companies that

---

[62] The democratic deficiency debate continues to stir scholarly attention. See Curtis
A. Bradley and Jack L. Goldsmith, "Customary International Law as Federal Common
Law: A Critique of the Modern Position" (1997) 110 Harv. L. Rev. 815; Curtis A. Bradley
and Jack L. Goldsmith, "The Current Illegitimacy of International Human Rights
Litigation" (1997) 66 Fordham L. Rev. 319; Phillip Trimble, "A Revisionist View of
Customary International Law" (1986) 33 UCLA L. Rev. 665; and the responses of Harold
Hongju Koh, "Commentary: Is International Law Really State Law?" (1998) 111 Harv.
L. Rev. 1824; and Gerald L. Neuman, "Sense and Nonsense about Customary
International Law: A Response to Professors Bradley and Goldsmith" (1997) 66
Fordham L. Rev. 371. On the probable influence of the democracy argument on the
courts, see Lea Brilmayer, "International Law in American Courts: A Modest Proposal"
(1991) 100 Yale LJ 2277; Richard A. Falk, *The Role of Domestic courts in the International
Legal Order* (Syracuse University Press, 1964); Thomas M. Franck, "The Courts, the State
Department, and National Policy: A Criterion for Judicial Abdication" (1960) 44 Minn.
L. Rev. 1101.

[63] On the importance of escape mechanisms for domestic interest groups, see the next
section below.

operate abroad. A judicial assertion, for example, that state parties to a bankrupt international organization are responsible for its outstanding obligations towards third parties (for example, that the state parties to the bankrupt London-based International Tin Council were responsible towards individual debtors for its debts)[64] might render enforcement only against the forum state (and not against the other state parties) or lead other organizations to seek a more accommodating seat. Enforcing strict domestic standards on domestic actors operating abroad – such as Union Carbide operating in India, for the Bhopal disaster,[65] or Texaco in South America, for polluting rivers in Ecuador and Peru[66] – may prompt those actors and others to leave that jurisdiction and set up office in the same jurisdiction as their competitors. National courts, just as national legislatures and governments do, understand this and join in the race to the bottom. The courts behave like any other actor in a prisoner's dilemma situation, pursuing one dominant strategy: the protection of domestic interests. Only when international guarantees provide judges with assurances that *all* their foreign colleagues will cooperate can this timidity be overcome. Such a guarantee is provided by, for example, Article 177 of the Treaty of Rome, which eliminates the possibility that the interpretation and implementation of Community law by national courts will diverge from that adopted by the central organ, the European

---

[64] In the wake of the collapse of the International Tin Council, claims of individual debtors were rejected owing to the immunity enjoyed by the organization. See *J. H. Rayner Ltd.* v. *Dep't of Trade & Indus.*, [1989] 3 Weekly Law Reports 969, 81 Int. L.R. 670 (HL); see also *Arab Org. for Industrialization* v. *Westland Helicopters Ltd.*, 80 Int. L.R. 622 (Fed. Sup. Ct. 1989) (Switzerland) (finding the insolvent AOI legally distinct from the state parties and hence finding the latter not liable for the AOI obligations). On this matter, see Michael Singer, "Jurisdictional Immunity of International Organizations: Human Rights and Functional Necessity Concerns" (1996) 36 Va. J. Int'l L. 53; Romana Sadurska and Christine M. Chinkin, "The Collapse of the International Tin Council: A Case of State Responsibility?" (1990) 30 Va. J. Int'l L. 845.

[65] *In re Union Carbide Corp. Gas Plant Disaster at Bhopal, India* in December 1984, 634 F. Supp. 842 (SDNY 1986), modified, 809 F.2d 195 (2d Cir. 1986), *cert. den.*, 484 US 871 (1987) (dismissed on the basis of *forum non conveniens*). Despite the Indian government's assertion that Indian Courts were unable to cope with the magnitude of these claims, the US Court ruled that litigating in India would respect Indian sovereignty and judicial self-sufficiency (634 F. Supp. at 867). The real underlying issue was, of course, the possibility of obtaining the higher US standards of damages, including punitive damages.

[66] *Sequihua* v. *Texaco* Inc., 847 F. Supp. 61 (S. D. Tex. 1994); *Aquinda* v. *Texaco Inc.*, 945 F. Supp. 625 (SDNY, 1996) (dismissed on grounds of *forum non conveniens* and international comity).

Court of Justice.[67] Only under such a regime will we see bold judicial enforcement of supranational norms, such as the implementation of the European Community's Habitats Directive.[68] Until this is ensured, domestic defectors can count on compliant national courts.

This thesis extends also to courts of regional organizations such as the European Court of Justice, the judicial organ of the European Union. Such courts tend to adopt a similar partisan attitude when dealing with the external relations of the organization. And indeed, the European Court of Justice has dismissed a number of challenges brought against the European Commission's and Council's discriminatory policies in relation to the "Banana War" involving the EU, Central American countries, and the US.[69]

A voice in international law-making

The disparity between the influence of small and large groups in the conclusion of international treaties is expressed not only in domestic processes and regional negotiations among governments of neighboring states. It is also manifested in international lawmaking via multilateral negotiations and in the emergence of norms through the mysterious process of custom creation, whereby a state that violates the prevailing custom may prove, in retrospect, to be the harbinger of the new law.[70] Although NGOs are increasingly finding ways to present their cases during the course of high-profile international negotiations for multilateral

[67] See Eyal Benvenisti, "Judges and Foreign Affairs: A Comment on the Resolution of the Institute of International Law on 'National Courts and the International Relations of their State'" (1994) 5 Eur. J. Int'l L. 423.

[68] Directive 92/43/EEC on the Conservation of Natural Habitats and of Wild Flora and Fauna. For cases dealing with the interpretation of the Directive, see *R. v. Secretary of State for Trade & Industry, ex parte Greenpeace Ltd.* (QB, UK) (1999) (Lexis); *Swan v. Secretary of State for Scotland* (Court of Session: Outer House, UK) (1999) (Lexis); Re WWF – UK, (Court of Session: Outer House, UK) (1998) (Lexis).

[69] C-73/97 P, *French Republic v. Comafrica SpA & Dole Fresh Fruit Europe Ltd. & Co.*, (http://curia.eu.int/jurisp) (21 Jan. 1999); *Germany v. Council of the European Union*, [1994] ECR I-4973 (reprinted in (1995) 34 ILM 154). For criticism of this case (and the case of Chiquita Italia, [1996] 7 EuZW 118), see Hilf, "New Frontiers," note 33. But see *Germany & Belgium v. Council of the European Union*, http://curia.eu.int/jurisp (10 Mar. 1998) (annulling part of the Council's decision approving the conclusion of a trade agreement with third party states because it discriminates between different exporters of bananas to the EU); note, however, that at the time of this judgment, the relevant agreement was under challenge at the WTO (see note 50 and accompanying text).

[70] On this evolution of customary norms see chapter 8.

lawmaking agreements, especially in the environment sphere,[71] actual bargaining is left entirely to state representatives. NGOs usually do relatively better at high-profile, specific, and relative short-term occasions such as the Earth Summit in 1992 and the Rome Conference in 1998,[72] but are less effective at monitoring lengthy processes that receive less media attention, such as the debates at the International Law Commission. The documents they help produce – declarations and resolutions – bear no immediate legal significance, although in recent years, they have come to be regarded as "soft law," a somewhat dubious status for not-really-law that may influence governments in the often meandering course of the development of customs. As Christine Chinkin observes in referring to the output of NGO participation in the human rights area, "Although NGOs have made significant inroads, States retain a tight grip on the formal law-making process while apparently ceding ground."[73]

The process of international lawmaking in the context of shared freshwater is a telling case in point. Throughout the twentieth century, there were three parallel efforts to codify or, more precisely, formulate a law on international fresh water. The first attempt was made by the Institut de droit international (IDI), a prestigious private body comprised of international legal experts who serve at the Institute in a personal capacity. In 1911, the IDI set the frame of inquiry with its Madrid Resolution.[74] In 1961, it updated its position with the Salzburg Resolution.[75] The International Law Association (ILA), another highly respected and learned society open to all international lawyers through branch offices in many countries, has been engaged with this issue since 1956, and its resolutions – most notably, the Helsinki Rules of 1966 – are widely perceived to reflect customary international law.[76] Finally, the International

---

[71] See, e.g., Barbara J. Bramble and Gareth Porter, "Non-Governmental Organizations and the Making of US International Environmental Policy" in Andrew Hurrell and Benedict Kingsbury (eds.), *The International Politics of the Environment* (Oxford, Clarendon Press, 1992), p. 313; A. Dan Tarlock, "The Role of Non-Governmental Organizations in the Development of International Environmental Law" (1993) 68 Chi. Kent L. Rev. 61; Jeffrey L. Dunoff, "From Green to Global: Toward the Transformation of International Environmental Law" (1995) 19 Harv. Envtl. L. Rev. 241.

[72] See note 91 and accompanying text.

[73] Christine Chinkin, "Human Rights and the Politics of Representation: Is there a Role for International Law?" in Michael Byers (ed.), *The Role of Law in International Politics* (Oxford, Oxford University Press, 2000), p. 131 at 140.

[74] 24 Ann. Inst. Dr. Int'l 365 (1911).

[75] *Resolution on the Utilization of Non-Maritime International Waters (Except for Navigation) Adopted at its Session at Salzburg* (3–12 Sept. 1961), Article 6. See (1961) 49 (II) Ann. Inst. Dr. Int'l 370 (trans. in [1962] AJIL 737).

[76] On the development of the law, see chapter 7.

Law Commission (ILC), a UN body established by a General Assembly resolution and consisting of state representatives, was devoted to this matter
from 1971 until 1994, producing a set of draft articles for the United
Nations for adoption as a multilateral convention. Unlike the more
academic environments at the IDI and ILA, the mandate of the ILC,
the composition of its members, and the procedures it adopted all
ensure that governmental interests set the agenda (although, as will
be mentioned later, informal relationships with the IDI and, especially, the ILA do contribute to the discussions). The ILC Statute, which
assigns the objective of "the promotion of the progressive development
of international law and its codification,"[77] instructs the ILC to follow
the prevailing positive norms and reflect governments' interests. Where
state practice, precedent, and doctrine are already extensive, codification
would consist of a "more precise formulation and systematization" of the
prevailing law.[78] Where practice is not yet sufficiently developed or not
yet regulated by international law, the ILC is expected to prepare draft
conventions[79] in a procedure set forth by the Statute.[80] This procedure
involves input from the governments of all UN member states, through
responses to questionnaires and comments on drafts. State input is made
also during the discussion of the ILC work at the UN Sixth Committee
and, finally, in the conference convoked to conclude a convention on
the basis of the ILC work.

ILC members, all of whom must be "persons of recognized competence
in international law,"[81] are elected by the General Assembly from a list of
candidates nominated by the governments of the member states.[82] The
Commission should, as a whole, represent "the main forms of civilization and the principal legal systems of the world."[83] In fact, however,
there were concerns that "elections to the Commission have become
overpoliticized."[84] Together with the rigid key for geographical distribution of seats that has been adopted,[85] such practices were seen as
responsible for the composition of the Commission "reflect[ing] rather

---

[77] Statute of the International Law Commission, Article 1(1). See also Oscar Schachter,
International Law in Theory and Practice (Boston, M. Nijhoff, 1991), chapter 5; Herbert
W. Briggs, *The International Law Commission* (Ithaca, NY, Cornell University Press, 1965),
pp. 129–41; see also Robert Y. Jennings, "The Progressive Development of International
Law and its Codification" (1947) 24 Brit. YB Int'l L. 301.

[78] Statute of the International Law Commission, Article 15.

[79] *Ibid.*   [80] *Ibid.* Article 16.   [81] *Ibid.* Article 2(1).

[82] *Ibid.* Article 3.   [83] *Ibid.* Article 8.

[84] Sir Ian Sinclair, *The International Law Commission* (Cambridge, Grotius, 1988), p. 14.

[85] Sinclair, *The International Law Commission*, p. 15.

closely the political forces at the possible expense of the professional forces."[86]

The ILC's Statute allows it to "consult with scientific institutions and individual experts; these experts need not necessarily be nationals of Members of the United Nations";[87] and even to "consult with any international or national organization, official or non-official, on any subject entrusted to it if it believes that such a procedure might aid it in the performance of its functions."[88] But in practice, the "feeling in the Commission...has in general not been favourable to consultation with national official organizations and non-governmental organizations."[89] ILC members consulted informally with members of the ILA and the IDI, and at times, formal consultations were even held with the ILA.[90] As a result, NGOs only have an opportunity to become involved at a relatively late stage, if and when an international conference is convened to formulate a treaty on the basis of the ILC proposals. Despite this relatively late-stage involvement, NGOs can still play a significant role. The NGOs' potential was demonstrated during the Rome Conference in 1998, convened to formulate the Statute of the International Criminal Court. As Mahnoush Arsanjani describes,

[NGOs'] influence was felt on a variety of issues, particularly the protection of children, sexual violence, forced pregnancy, enforced sterilization and an independent role for the prosecutor. Throughout the Preparatory Committee's sessions and the Rome Conference, they provided briefings and legal memoranda for sympathetic delegations, approached delegations to discuss their points of view, and even assigned legal interns to small delegations. On occasion, they increased pressure on unsympathetic delegations by listing them as such in the media.[91]

This account highlights the crucial role NGOs can play in the international law-making process, promoting the concerns of the unprivileged and those with little political clout.

---

[86] Shabtai Rosenne, "The International Law Commission, 1949–1959" (1960) 36 Brit. YB. Int'l L. 107 at 130; Sinclair, *The International Law Commission*, note 84, at pp. 16–17.

[87] See note 77, Article 16(e).   [88] *Ibid.* Article 26(1).

[89] See generally "The Work of International Law Commission" (5th edn, New York, United Nations, 1996), pp. 13–21. The above quote is taken from the fourth (1988) edition of the book, at p. 24.

[90] Subir Goswami, *Politics in Law Making* (New Delhi, Ashish Publishing House, 1986), p. 197. The only other body recognized by the ILC as a potential consulting body has been the IDI. See (1965) 2 *ILC Yearbook*, at 194–5 para. 64(c).

[91] Mahnoush H. Arsanjani, "The Rome Statute of the International Criminal Court" (1999) 93 AJIL 22 at 23.

As a result of the ILC composition and procedures, the law-making process is dominated by lawyers serving as the representatives of the different governments. While the ILA and IDI do not necessarily consist only of governments' representatives, they do include only international lawyers. Both, and in particular, the more conservative IDI, are not inclined to venture anywhere beyond a conservative study and interpretation of state practice and of treaties, nor do they seek to devise innovative proposals. As Christine Chinkin writes, "It is an irony of contemporary international law-making that it is in fact the [UN] General Assembly itself that is more likely to advance the progressive development of the law...whereas the work of the [ILC] may in fact serve to inhibit further evolution of the law."[92] Although these private bodies are better able to represent diverse, not necessarily governmental interests, their methods of research and analysis remain limited. This disadvantage is more apparent when the subject-matter departs from the bread-and-butter, typical lawyers' law and requires a sophisticated understanding of such spheres as hydrology and geology.

It should come as no surprise, then, that the results of such a legislative process, shepherded by a group of lawyers and strongly influenced by those among them who represent the interests of incumbent governments, conform to the expectations of those governments. Even more dramatic, the exclusion of extra-legal, scientific evidence prevents proper analysis of pertinent issues and plays into the hands of parties who benefit from the absence of such an analysis. Take as a glaring example the ILC decision in 1994, after two decades of deliberation, to exclude transboundary "confined aquifers" (namely, aquifers whose waters do not contribute to a system that includes water above ground) from the regime of what later became the 1997 United Nations Convention on the Law of the Non-Navigational Uses of International Watercourses ("the Watercourses Convention").[93] Such aquifers raise the same collective-action problems as surface water does and therefore merit and,

---

[92] Christine Chinkin, "Enhancing the International Law Commission's Relationships with Other Law-Making Bodies and Relevant Academic and Professional Institutions" in *Making Better International Law: The International Law Commission* (United Nations, 1998), p. 50, at pp. 333, 336.

[93] See the declaration on "confined aquifers," appended to the 1994 Draft Articles (*Report of the ILC on the Work of its Forty-sixth Session*, UNGAOR 49th Session Supp. No.10 [A/49/10] [1994], 195, at p. 326). The United Nations Convention on the Law of the Non-Navigational Uses of International Watercourses, adopted 21 May 1997, reprinted in (1997) 36 ILM 700.

indeed, have received on many occasions the same treatment.[94] The ILC, however, chose not to include these aquifers in the definition of shared water resources ("a watercourse system" in the terminology of the ILC), because many members felt that the nature of these aquifers must be further studied before addressing their use. Instead, the ILC adopted a resolution recommending that states regulating transboundary ground-water be guided by the draft articles, where appropriate.[95] Had simple hydrologic facts been sought from non-legal experts at an early stage in the process, members of the ILC would not have been able to claim that the characteristics of such aquifers remain too vague to be covered by the Convention.

Stock taking

The making, shaping, and breaking of international commitments – through bilateral and multilateral treaties, through customs, and through individual acts of breach – are the domain of national governments. In terms of citizens' opportunities to influence the outcomes, these prescriptive efforts are characterized by small-group domination, with a dearth of formal and informal opportunities for the larger public to participate.[96] This ensures small interest groups a significant edge over other domestic groups in securing their sectarian interests. It contributes to welfare and even efficiency losses and to the growing democratic deficit identified in the previous chapter. If the Westphalian

---

[94] On the need to regulate confined aquifers under the same principles, see Special Rapporteur Robert Rosenstock's *Second Report on the Law of the Non-Navigational Uses of International Watercourses*, A/CN.4/462, 21 April 1994, Annex (pp. 22–35); the ILA's Seoul Rules [Report on The Sixty-Second Conference, 251 (1987)], Article 1 and comment to Article 1, at 251–8); ECE, Charter on Groundwater Management, ECE Annual Report (1989–90), ECOSOCOR 1989, Supp. No. 15; 1992 Helsinki Convention on the Protection and Use of Transboundary Watercourses and International Lakes, reprinted in (1992) 31 ILM 1312, Article 1(1); Robert D. Hayton and Albert E. Utton, "Transboundary Groundwater: The Bellagio Draft Treaty" (1989) 29 Nat. Res. J. 663; Dante Caponera and Dominique Alheritiere, "Principles for International Groundwater Law" (1978) 18 Nat. Res. J. 589; Julio Barberis, "The Development of International Law of Transboundary Groundwater" (1991) 31 Nat. Res. J. 167.

[95] *Report of the ILC on the Work of Its Forty-sixth Session*, UNGAOR 49th Session Supp. No. 10 (A/49/10) (1994), 195, at p. 326.

[96] Recently, it was recommended that the ILC consult NGOs on a regular basis. See Chinkin, "Enhancing the ILC," note 92, at pp. 340–1; *cf.* discussion at in *Report of the ILC* at pp. 134–50 and recommendations at 42–3. However, skepticism was also voiced concerning the possibility that the consumers of the product of ILC deliberations, namely, governments, would be amenable to increasing NGO influence (see, for example, the comments made by John Dugard, at p. 147).

paradigm and the national and international structures that support it fail on these crucial fronts, can it be amended and how? My attempt to respond to this challenge is twofold: I suggest new institutional designs and identify legal norms that could level the political playing field where small groups compete with larger ones over political influence and could provide for equitable and sustainable collective action.

Transnational institutions as a response

The transnational conflict paradigm described in the previous chapter suggests that global market failures are often the result of transnational conflict among interest groups rather than international conflict among unitary states. As described in the first part of this chapter, the Westphalian-based legal and political system – both domestic and international – ignores this reality and only serves to perpetuate inter-group externalization of costs and, consequently, inefficient, non-sustainable, and inequitable outcomes. Moreover, it undermines the idea of democratic governance.

This part suggests that new tools must be designed to accommodate the true nature of those transnational conflicts that dominate the transnational conflict paradigm. It calls for a new conception of the opportunities provided by regional and international institutions as instruments of management of transnational conflicts. Opportunities for public participation in the decision-making processes within these institutions would provide an effective voice for the public and a new meaning to democratic participation in the management of the global commons.

A definition

A more structured decision-making process could limit significantly the opportunities for small domestic interest groups, bureaucrats, and politicians to pursue short-term sectarian goals to the detriment of society at large and future generations. My claim is that transnational institutions would be capable of responding to a great number of global collective-action problems in ways that promote not only efficiency, but also democracy and social justice. With these purposes in mind, and for reasons to be explained below, I define transnational institutions as treaty-based procedures for the coordination of policies with respect to a specific activity (such as trade, taxation, or anti-trust) or a specific shared resource. Such institutions would include, at the very least, permanent bodies and permanent processes for the collection of relevant data, their assessment and dissemination to the public, and the formulation of publicly stated and reasoned policies (in the form of opinions, recommendations, or decisions) on the basis of those data. As part of their data processing, such institutions should implement mechanisms for monitoring compliance by the various domestic actors with the institutions' policies, amongst other things, by providing access to and soliciting input from NGOs representing different interest groups. Such mechanisms will reduce informational asymmetries and allow a more meaningful role for representatives of the larger domestic interest groups, such as consumers, employees, and environmentalists. In short, transnational institutions would mirror the better[97] types of contemporary, domestic administrative agencies on the global level, while preserving and respecting the still-cherished principle of state sovereignty.

My definition of transnational institutions is narrower than that adopted by international relations theorists,[98] as it relies upon the decision-making process as a tool for reducing uncertainties and allows a more meaningful role for representatives of larger domestic groups. As explained in the next section, transnational institutions under my definition would provide a forum for the evolution of conditional cooperation that can resolve collective-action problems.

These transnational institutions would not be intended to replace governments in implementing institutional prescriptions. They would respect the principle of state sovereignty. State control would be retained over decision making through the requirement of a consensus amongst participating states or, in extreme cases, by means of the exit option. Moreover, implementation of the institutions' policies could, in

---

[97] For an analysis of novel decision-making processes within administrative agencies in the US see Michael C. Dorf and Charles F. Sabel, "A Constitution of Democratic Experimentalism" (1998) 98 Colum. L. Rev. 267.

[98] According to Stephen D. Krasner's widely accepted, albeit rather loose, definition, international institutions are "sets of implicit or explicit principles, norms, rules, and decision-making procedures around which actors' expectations converge in a given area of international relations." Stephan D. Krasner (ed.), *International Regimes*, (Ithaca, NY, Cornell University Press, 1983). On different types of international institutions, see Oran R. Young, *International Cooperation* (Ithaca, NY, Cornell University Press, 1989). For a categorization of regimes according to the process through which they are established (spontaneous, negotiated, imposed), see Oran R. Young, "The Rise and Fall of International Regimes" (1982) 36 Int'l Org. 277. See also Stephan Haggard and Beth A. Simmons, "Theories of International Regimes" (1987) 41 Int'l Org. 491 at 493–6.

most cases, be left in the hands of the states (although domestic control mechanisms would be subject to scrutiny by the institutions). The
transnational institutions could have autonomous tools for sanctioning
violators, but independent policing powers would not be a prerequisite
for the effectiveness of the institutions. Their provision of information, formal and accessible decision-making procedures, and monitoring would be sufficient for ensuring an efficient and more democratic
allocation of competences and resources.

Institutions and the emergence of collective action: a general theory

The neo-liberal school of international relations has distinguished itself
by emphasizing the role of institutions in international politics. As explained by Robert Keohane, one of this school's leading theoreticians,
the effectiveness of international institutions lies in the establishment
of frameworks for exchange of information, mutual monitoring, and
frequent interaction. Such a forum encourages the development of stable mutual expectations regarding the future behavior of co-parties and
reduces both bargaining costs and uncertainty with regard to the value
of proposed transactions.[99] Such institutions appear "whenever the costs
of communication, monitoring, and enforcement are relatively low compared to the benefits to be derived from political exchange."[100]

   Critics, however, call into question the utility of transnational institutions, suggesting that these institutions, in being composed of
decision-makers from all participating states, merely mirror the typical
negotiating scenarios between state negotiators.[101] In what way, these
critics ask, do such institutions transform the regular interstate bargaining process and the domination by small interest groups of that process?
Proponents of the Westphalian paradigm have responded by suggesting that transnational institutions reduce the phenomenon of asymmetric information among members (in exposing free-riders) and other
transaction costs related to negotiating agreements and sanctioning

---

[99] See Robert O. Keohane, *After Hegemony* (Princeton, NJ, Princeton University Press,
     1984), pp. 85–98.
[100] Robert O. Keohane, *International Institutions and State Power* (Boulder, CO, Westview
     Press, 1989), p. 167.
[101] The debate concerning the effectiveness of IOs as opposed to negotiations has
     haunted IR scholars since the 1980s: see Lisa L. Martin and Beth A. Simmons,
     "Theories and Empirical Studies of International Institutions" (1998) 52 Int'l Org. 729
     at 742–7.

defectors; and they provide the legitimizing effects of a communal forum (in identifying outsiders as free-riders), benefiting from the role played by autonomous bureaucracy, scientific personnel, and sometimes also courts.[102] But I submit that institutions matter mainly because they exist in a political environment that is more aptly described by the transnational conflict paradigm. This paradigm provides a better insight into the understanding of international institutions. It indicates that these institutions are important not only in reducing inter-state transaction costs and enhancing communication. Rather, they transcend the veil of sovereignty and reduce capture by domestic small interest groups whose uses are often wasteful.[103] Transnational institutions, as defined above, transform a largely unstructured and veiled negotiations process, followed by an insufficiently informed ratification process where the deal is presented as a "take it or leave it" option, into a well-defined, widely accessible, and transparent decision-making procedure.[104] Scrutiny of this procedure is available throughout its elaboration; wide representation is allowed; and, thus, capture is reduced.

Deliberations by means of transnational institutions change the opportunities and relative control of different actors – representatives of various domestic interests, politicians, and bureaucrats – in the decision-making process.[105] The structured institutional decision-making process

---

[102] See Kenneth W. Abbott and Duncan Snidal, "Why States Act through Formal International Organizations" (1998) 42(1) J. Conflict Res. 3; Alexander Thompson, "Unilateral Enforcement in the Shadow of International Institutions: Canada, Spain, and the Northwest Atlantic Fisheries Organization" (paper prepared for presentation at the Annual Meeting of the American Political Science Association, September 1998) (discussing the legitimization effects of the NAFO decision and information-gathering on Canada's unilateral enforcement against Spain, exposed as a free-rider). Ever-closer interstate cooperation was explained by referring to the influence of "epistemic communities" (i.e., experts sharing common practices and policies: see Peter M. Haas (ed.), "Knowledge, Power, and International Policy Coordination" (1992) 46 Int'l Org. 1 (Special Issue)) and to the interaction within "government networks" (Anne-Marie Slaughter, "Governing the Global Economy through Government Networks" in Michael Byers (ed.), *The Role of Law in International Politics* (Oxford, Oxford University Press, 2000), p. 177).

[103] For an early assessment of the achievements of institutions in protecting the environment, see Peter M. Haas, Robert O. Keohane, and Marc A. Levy (eds.), *Institutions for the Earth* (Cambridge, MA, MIT Press, 1993).

[104] Witness the slow but steady move of EU institutions towards transparency, parallel to the enlargement of their powers and the need to address public concern with the democratic deficit: Craig and de Burca, *EU Law*, note 22, at pp. 368–71; Veerle Deckmyn and Ian Thomson (eds.), *Openness and Transparency in the European Union* (Maastricht, European Institute of Public Administration, 1998).

[105] For a similar claim with respect to NGO participation in WTO procedures, see Daniel C. Esty, "Non-Governmental Organizations at the World Trade Organization:

counterbalances the relative edge smaller domestic groups have in obtaining information and exerting leverage during the negotiations and ratification process. Information asymmetry is reduced by the dissemination of the data collected and assessed within the institution. Public participation is enhanced by the opportunities provided to representatives of the larger domestic groups to partake in the decision-making process, to comment on suggested policies, and to offer alternative plans. In the field of the environment, improved access to information and public participation in decision making enhance also the implementation of decisions, as they contribute to public awareness of environmental issues and strengthen public support for decisions on the environment.[106] Elimination of the ratification process reduces the potential of smaller groups to concentrate their capture efforts in the national institutions. Public scrutiny, rather than coming into play only at the stage of ratification of the decision of a transnational institution, will focus instead on the institution's stated reasons for its decision. The institutions, in contrast to negotiators and in close resemblance to administrative agencies or even courts, can, and should, state their reasons for the actions they take.

This process will not only serve to alter the relative leverage of the conflicting groups; the personal composition of the bureaucracy of the institutions can contribute significantly to an environment that permits less capture. Because the structured decision-making process is supposed to rely on the accumulation and assessment of data, decision making would involve less-politicized personnel. Scientists would be processing the data, thereby potentially providing common ground that politicians cannot ignore in their deliberations regarding the institution's policies. Although scientists may differ in opinions, and the uncertainties pointed out by their research could be exploited for political ends,[107] there is solid evidence that their involvement reduces conflicts

---

Cooperation, Competition, or Exclusion" (1998) 1 J. of Int'l Econ. L. 123; Daniel C. Esty "Linkages and Governance: NGOs at the World Trade Organization" (1998) 19 U Pa. J. Int'l Econ. L. 709.

[106] These points are mentioned in the preamble to the Convention on Access to Information, Public Participation in Decision-Making and Access to Justice in Environmental Matters, adopted in Aarhus, Denmark on 25 June 1998 by member states of the Economic Commission for Europe and other European states (reprinted in (1999) 38 ILM 517).

[107] Ostensibly "scientific" arguments can sometimes be exploited to support non-cooperative policies. On scientific uncertainties and value judgments in the context of the sustainable management of the environment, see John Lemons and Donald A. Brown (eds.), *Sustainable Development: Science, Ethics, and Public Policy*

and improves the likelihood of international cooperation.[108] Ecosystem management relies, to a large extent, on scientific research on supply management and on engineering input with respect to the design of water and other installations, emission controls, and irrigation systems. The exchange of such apolitical data, mainly amongst experts and low-level officials, would contribute to providing a less-politicized shared database, which would constrain the range of options open to political actors. As described by David LeMarquand, who observed the work of the US–Canada International Joint Commission, "informal intelligence gathering [by low-level officials] help[ed] to provide an early warning of impending issues, and permit[ted] actions before issues [became] too politicized."[109] This scientific work provides common ground that cannot be disregarded by governments in subsequent negotiations; it raises the level of concern about threats to the common resource and indicates the policies appropriate for addressing these threats. Finally, these institutions are a source for disseminating information to the domestic public in the participating states and a forum that enables NGOs to take an active role in the decision-making process. Domestic public pressure has proved to be a key factor in changing governmental policies: public interest in environment protection and sustainable use can outweigh agricultural or industrial lobbies that push for short-term goals. Institutions that have opened channels to domestic public opinion have

(Dordrecht, Kluwer Academic Publishers, 1995). On the use of science to legitimize political interests, see Yaron Ezrahi, *The Descent of Icarus* (Cambridge, MA, Harvard University Press, 1990), pp. 210–36; Wendy E. Wagner, "The Science Charade in Toxic Risk Regulation" (1995) 95 Colum. L. Rev. 1613. On the use and misuse of science-based risk assessment in support of states' adoption of strict trade policies under the WTO 1994 Agreement on Sanitary and Phytosanitary Measures, see Robert Howse, "Democracy, Science and Free Trade: Risk Regulation on Trial at the World Trade Organization" (unpublished manuscript, on file with author, 1999).

[108] This has proven crucial even in the context of Canada–US relations: David G. LeMarquand, "Preconditions to Cooperation in Canada–United States Boundary Waters" (1986) 26 Nat. Res. J. 221 at 232. For an analysis of the Canada–US International Joint Commission, see notes 108, 109, 114, 115 and accompanying text. This has also been the experience in the development of the Antarctic Treaty System: Lorraine M. Elliott, *International Environmental Politics: Protecting the Antarctic* (New York, St. Martin's Press, 1994), pp. 43–4; Jonathan Blum, "The Deep Freeze: Torts, Choice of Law, and the Antarctic Treaty System: Is It Adequate to Regulate or Eliminate the Environmental Exploitation of the Globe's Last Wilderness?" (1992) 14 Hous. J. Int'l L. 597 at 673.

[109] David G. LeMarquand, "Preconditions to Cooperation" note 108, at 232.

succeeded in bringing about difficult policy changes in the participating states. As Haas *et al.* observed, when examining the relative success of international institutions, "If there is one key variable accounting for policy change, it is the degree of domestic pressure in major industrialized democracies, not the decision-making rules of the relevant international institution."[110]

Transnational institutions would have their own independent bureaucracy, which would, therefore, identify with the institution's success and reputation as its own. Such bureaucrats will try to extend their powers and will develop an innate tendency towards more intensified cooperation with one another.[111] The information disseminated to the general public will constrain the range of options open to the national politicians working inside the institution and operate as a check on the decisions they make.

Interaction among transnational institutions entrusted with overlapping or tangential matters – water and forests management or environment and trade regimes are only two examples – would prove beneficial to the goals of efficient and sustainable collective action. Such a novel modality of transnational checks and balances could create inter-institution friction and a consequent increase in opportunities for public scrutiny and accountability of decision-makers.

The information-inviting and -generating mechanisms will provide a voice for all affected interest groups. As the literature on the emergence of cooperation in the management of common-property resources suggests, institutions that provide for equal voice are likely to resolve the collective-action problem they must contend with.[112] In addition to voice, such institutions would be able to hold domestic officials accountable for their acts and omissions that affect the transnational subject-matter. They could draw the public's attention to ineffective domestic regulation of private activities.

---

[110] Haas *et al.*, *Institutions*, note 103, at 14. See also pp. 399–400.

[111] See, for example, the recent US–Canada IJC proposal, *The IJC and the 21st Century*, suggesting the establishment of permanent IJC international watershed boards to manage additional major transboundary basins (Press Release, 23 Nov. 1998, http://www.ijc.org/news/h2oshed1198.html).

[112] See Elinor Ostrom, *Governing the Commons* (New York, Cambridge University Press, 1990), chapter 6. For examples of institutions successfully providing voice to both producers and consumers of water, see Barton H. Thompson, Jr., "Institutional Perspectives on Water Policy and Markets" (1993) 81 Calif. L. Rev. 671 at 687–93. On successful water institutions in ancient societies, see chapter 1.

For these reasons, it is suggested that the efficacy of these institutions should not rely on independent enforcement powers. Given the transnational nature of the conflicts, enforcement should remain the private initiative of the non-state actors acting within the domestic polity and using the national political and judicial fora. Well-informed NGOs could, for example, respond to free-riding attempts by other sub-state actors (and governmental compliance with such attempts) by waging domestic public opinion campaigns, thereby exerting effective political pressure on incumbent governments and members of the legislatures. They could also, although probably with less success,[113] petition the domestic courts for judicial review. In short, they would be able to act no less effectively than when they respond to entirely domestic policies that affect the interests of those they represent.

One example of effective regional collective action is the US–Canada International Joint Commission, entrusted with the management of the rivers shared by the two states.[114] Of the 110 applications for decisions submitted to the Commission up until 1987, only in four cases did some of the members give dissenting opinions, and only in two of these cases did the dissents follow national policy lines. Moreover, both the US and Canadian governments adopted more than three-quarters of the total decisions.[115] Another example is the Finnish–Swedish Frontier Rivers Commission established in 1977, whose decisions have always been unanimously accepted, despite the fact that only a majority vote was necessary for passing a decision.[116] For both these Commissions, information and voice were sufficient substitutes for independent enforcement measures.

An example of a global organization whose effectiveness is enhanced by wide participation is the International Labour Organization. Deliberations follow the concept of "tripartism": each state's delegation to all the deliberative bodies is comprised of two government representatives, one representative of employers, and one representative of workers.

---

[113] See the discussion in notes 14–33 and accompanying text, on the reluctance of courts to interfere in issues that constrain governmental activities beyond national borders.

[114] On the IJC, see Stephen J. Toope and Jutta Brunnée, "Freshwater Regimes: The Mandate of the International Joint Commission" (1998) 15 Ariz. J. Int'l & Comp. Law 273; Patricia K. Wouters, "Allocation of the Non-Navigational Uses of International Watercourses: Efforts at Codification and the Experience of Canada and the United States" (1992) 30 Can. Yb. Int'l L. 43.

[115] Wouters, "Allocation," note 114, at pp. 78–9 n. 186 (citing R. B. Bilder's Working Paper, *When Neighbours Quarrel: Canada–US Dispute Settlement Experience* (1987)).

[116] Malgosia Fitzmaurice, "The Finnish–Swedish Frontier Rivers Commission" (1992) 5 Hague YB. Int'l L. 33 at 44.

The representatives of the employers and workers are instrumental in negotiating the labor standards and in implementing them within the national jurisdictions.[117]

This is not to suggest that transnational institutions would not face challenges to their authority. In states where small interest groups enjoy relatively high political leverage, one can expect less commitment to compliance. Deep-seated animosities among rival communities could overcome rational considerations and hinder cooperation. But it can be expected that as collective-action failures become more dramatic and their repercussions are fully grasped by the wider segments of the respective societies, popular demand will eventually produce such institutions.

Transnational institutions and democracy

The structured decision-making process in transnational institutions not only provides mechanisms to prevent market failures, especially in transboundary resources, it also responds to the democratic challenge, by providing three essential democratic tools. First, and perhaps most importantly, the improved information that these institutions provide to the voter counterbalances voters' inherent deficiency in monitoring the domestic institutions' performance. Both the domestic policies and the international engagements of the government are exposed to voter scrutiny. Such information will not only contribute to enhanced accountability of the state officials:[118] by informing the voters about their government's performance, transnational institutions offer them a more meaningful voice. Thus, these institutions enhance domestic democracy.

Second, the open access to the institutions' procedures provides domestic populations with the opportunity to influence regional and

---

[117] See Hector Bartolomei de la Cruz *et al.*, *The International Labor Organization: The International Standards System and Basic Human Rights* (Boulder, CO, Westview Press, 1996), p. 10 ("Tripartism is the real strength of the ILO"); see also Klaus Samson, "The Standard-Setting and Supervisory System of the International Labour Organisation" in Raija Hanski and Markku Suksi (eds.), *An Introduction to the International Protection of Human Rights* (Turku/Abo, Institute for Human Rights, Abo Akademi University, 1997), p. 149 at pp. 151–2.

[118] See Susanne Lohmann, "An Information Rationale for the Power of Special Interests" (1998) 92(4) Am. Pol. Sci. Rev. 809; see also Michael D. Rosenbaum, "Domestic Bureaucracies and the International Trade Regime: The Law and Economics of Administrative Law and Administratively-Imposed Trade Barriers" (Discussion Paper No. 250 1/99, The Center for Law, Economics and Business, Harvard Law School), available at http://www.law.harvard.edu/programs /olin_center (arguing that administrative procedures that lower the costs of access to information shift power over policy-making from more organized to less organized groups).

global policies that affect their well-being and choices. This access provides a voice without a vote. Cynics might criticize such an imperfect voice as being politically meaningless, and indeed this could be the case. But when strong cross-cultural links exist between communities,[119] voices will matter even when no vote supports them. A strong belief in the effectiveness of such voice was the reason given by the German Constitutional Court in its approval of Germany's ratification of the Maastricht Treaty.[120] The court expressed its belief in an integrated European Union, where there is an "ongoing free interaction of social forces, interests, and ideas, in the course of which political objectives are also clarified and modified, and as a result of which public opinion moulds political policy."[121] But for this to be achieved, according to the court,

it is essential that both the decision-making process amongst those institutions which implement sovereign power and the political objectives in each case should be clear and comprehensible to all, and also that the enfranchised citizen should be able to use its own language in communicating with the sovereign power to which it is subject.[122]

While this is a reflection of the court's assessment of the political environment in the European Union, it might also reflect emerging concern with the deliberative process that precedes the actual taking of votes. Proponents of a voice without vote argue that the process of reasoning and persuasion that precedes the actual vote is effective in eliminating Pareto-inferior outcomes by providing for more equitable distribution of resources in ways that reflect community goals and values. Such a process would also legitimize the decision taken, and thus, more actors would comply with it.[123] Transnational institutions could become forums of democratic deliberation and, in any event, would enrich the domestic deliberation processes in the participating countries. The duty of transnational institutions to provide reasons for their decisions will increase the accountability of the decision-makers, just as the reasoning given in court opinions serves as constraint on judicial power.

[119] See Anne-Marie Slaughter, "International Law in a World of Liberal States" (1995) 6 Eur. J. Int'l L. 503.
[120] Federal Constitutional Court Decision concerning the Maastricht Treaty of 12 October 1993 (translated in [1994] 33 ILM 388), at 420 (citations omitted).
[121] *Ibid.*   [122] *Ibid.*
[123] See Juergen Habermas, *The Theory of Communicative Action* (Boston, Beacon Press, vol. I, 1984; vol. II, 1987); Jon Elster (ed.), *Deliberative Democracy* (New York, Cambridge University Press, 1998).

The third democratic tool is the removal of specific matters from the sphere of influence of small and large interest groups. Venerable democratic institutions such as the courts and central banks are designed to be insulated from the immediate influence of voters and political pressures. To the extent that such insulation from politics is legitimate, transnational institutions can make it happen. Thus, for example, the recent European monetary integration, which entailed a complete loss of sovereign control over monetary policies for the member states, is compatible with democracy. In approving this transfer of powers, the German Constitutional Court accepted the government's "scientifically proven" argument that such a move was necessary to "ensure that the currency is not vulnerable to pressure groups or to holders of political office seeking re-election."[124] It is interesting to note that scientific findings have, in fact, suggested that such a transfer of power suits the interests of the business sector and governments, and hence, its adoption.[125] This notwithstanding, the removal of any matter from political control in a transnational institution will always have to be justified on the basis of democratic legitimacy.

Setting up institutions

The above discussion suggests that close attention must be paid to the procedural safeguards within transnational institutions. It also emerges that recognition must be given to the importance of the involvement of NGOs, the modern tribunes of the large domestic groups, in the decision-making process. The discussion explains why NGO participation is not a matter of charity, but a matter of right (the democratic right of participation in government) and of interest (of obtaining optimal outcomes without externality effects).

---

[124] See note 120, at 439. The court reasoned (at 439–40) that "this modification of the principle of democracy, which is designed to secure the confidence of making payment that is placed in a currency, is justifiable, because it takes account of the special factor, established in the German system and also scientifically proven, that an independent central bank is more likely to protect monetary value, and therefore the general economic basis for national budget policy and private planning and disposition, while maintaining economic liberty than are sovereign governmental institutions."

[125] See Geoffrey Garrett and Peter Lange, "Internationalization, Institutions and Political Change" in Robert O. Keohane and Helen V. Milner (eds.), *Internationalization and Domestic Politics* (New York, Cambridge University Press, 1996), p. 48 at pp. 66–9.

The setting-up of transnational institutions as well as the modifications in their composition and procedures are, in themselves, collective-
action problems, which could involve attempts to capture opportunistic
gains.[126] In this respect we can observe a conflict between governments,
especially from the developing countries, and NGOs on the proper role
of the latter in existing and designed institutions. There are even clear
examples of governmental efforts to repudiate outright the value of
these institutions. Thus, for example, only after violent demonstrations
did the decision-making bodies of the WTO become more accommodating towards increased transparency and NGO participation;[127] and the
1997 Watercourses Convention,[128] adopted after *decades* of negotiations,
failed to emphasize the importance of transnational institutions and
did not grant any role to NGOs, despite the insistence of some member
states and scholars.[129] A number of international lawyers have presented
scholarly support for the claims in favor of NGO intervention in international decision-making processes, basing these claims in the language
of civil and political rights.[130] But thus far, formal recognition of the

---

[126] On such gains within the international aviation markets, see John E. Richards,
"Towards a Positive Theory of International Institutions: Regulating International
Aviation Markets" (1999) 53 Int'l Org. 1 (demonstrating capture gains by domestic
groups in setting up international regulatory institutions in the international
aviation market).

[127] On the reluctance of the WTO in this respect and scholarly criticism of this
reluctance, see Esty, "NGOs," note 105; Arie Reich, "From Diplomacy to Law: The
Judicization of International Trade Relations" (1997) 17 J. Int'l L. Bus. 775 at 847–8
(discussing the democratization of the WTO procedures); Brian J. Schoenborn,
"Public Participation in Trade Negotiations: Open Agreements, Openly Arrived At?"
(1995) 4 Minn. J. Global Trade 103 (arguing in favor of increased public participation
in international trade negotiations); Philip M. Nichols, "Participation of
Nongovernmental Parties in the World Trade Organization: Extension of Standing in
World Trade Organization Disputes to Nongovernment Parties" (1996) 17 U. Pa. J. Int'l
Econ. L. 295 (discussing the issue of standing of non-governmental entities).

[128] Adopted on 21 May 1997, reprinted in (1997) 36 ILM 700.

[129] See the Report of Special Rapporteur Stephen S. McCaffrey, *Sixth Report on the Law of
the Non-Navigational Uses of International Watercourses*, (1990) UN. Doc. A/CN.4/427/Add.1.
For criticism, see Gunther Handl, "The International Law Commission's Draft Articles
on the Law of International Watercourses (General Principles and Planned Measures):
Progressive or Retrogressive Development of International Law?" (1992) 3 Colo. J. Int'l
Env. L. & Pol'y 123 at 124–9; Sergei V. Vinogradov, "Observations on the ILC's Draft
Rules: 'Management and Domestic Remedies'" (1992) 3 Colo. J. Int'l Env. L. & Pol'y
235 at 235–41; Constance D. Hunt, "Implementation: Joint Institutional Management
and Remedies in Domestic Tribunals (Articles 26–28 and 30–32)" (1992) 3 Colo. J. Int'l
Env. L. & Pol'y 281 at 284–7.

[130] See Phillip R. Trimble, "Globalization, International Institutions, and the Erosion of
National Sovereignty and Democracy" (1997) 95 Mich. L. Rev. 1944 at 1967–8; Alan
Boyle, "The Role of International Human Rights Law in the Protection of the

role and importance of NGOs in this capacity can be found only in non-binding international declarations, such as the 1992 Rio Declaration[131] and Agenda 21.[132]

Given the dominant position of smaller domestic interest groups in international lawmaking, the failure to set up adequate transnational institutions is not surprising. However, as soon as the larger domestic groups transcend the political boundaries, become aware of the stakes involved, and overcome the difficulties in coordinating their policies, pressure to reform the decision-making processes can be expected and must be accommodated.[133] The process of designing transnational institutions must include the participation of the wider public, indirectly through their representative NGOs and directly through the dissemination of user-friendly information. This design process will not be an easy task. A delicate balance will need to be found to accommodate governmental, inter-governmental, and non-governmental representation and to ensure that narrow interests, including those advanced by NGOs, do not gain dominance in the designing process.

Environment," in Alan E. Boyle and Michael R. Anderson (eds.), *Human Rights Approaches to Environmental Protection* (New York, Oxford University Press, 1996), p. 43 at p. 59; James Cameron and Ruth Mackenzie, "Access to Environmental Justice and Procedural Rights in International Institutions" in Boyle and Anderson, *Human Rights Approaches*, p. 129, at pp. 134–5; Sionaidh Douglas-Scott, "Environmental Rights in the European Union – Participatory Democracy or Democratic Deficit?" in Boyle and Anderson, *Human Rights Approaches*, p. 109, at pp. 112–20); Geoffrey Palmer, "New Ways to Make International Environmental Law" (1992) 86 AJIL 259.

[131] The 1992 Rio Declaration notes that "environmental issues are best handled with the participation of all concerned citizens, at the relevant level." (Declaration of the UN Conference on Environment and Development, Rio de Janeiro, 3–14 June 1992, Principle 10).

[132] Chapter 18 of Agenda 21 calls for active public participation in shared freshwater management, which includes not only the provision of a right of hearing to oppose plans that could be detrimental to certain individuals or groups, but, more generally, requires states to aim for "an approach of full public participation, including that of women, youth, indigenous people and local communities in water management policy-making and decision-making" (Agenda 21, Principle 18.9(c), reprinted in Nicholas A. Robinson (ed.), vol. IV, *Agenda 21 & the UNCED Proceedings* (New York, Oceana Publications, 1992–3), p. 57 at p. 66 and suggests the "development of public participatory techniques and their implementation in decision-making." (*Ibid.* Principle 18.12(n).) See also Ellen Hey, "Sustainable Use of Shared Water Resources: The Need for a Paradigmatic Shift in International Watercourses Law", in Gerald H. Blake *et al.* (eds.), *The Peaceful Management of Transboundary Resources* (London, Graham & Trotman, 1995), p. 127 at 133.

[133] One radical response to these challenges is the establishment of a "cosmopolitan democracy," based on an international constitution that provides for separation of different institutional powers and individual rights: David Held, *Democracy and the Global Order* (Stanford, CA, Stanford University Press, 1995), pp. 267–83.

The operative conclusion stemming from this analysis is that a pre-requisite to efficient, sustainable, and democratic management of trans-boundary resources requires setting up effective transnational institutions. Such an endeavor must respond to the particular characteristics of the resource or resources in question. If fresh waters are the focus of inquiry, institution builders must consider the diverse and conflicting uses of the resource. In ancient times, successful cooperation in water management was characterized by a lack of conflict over the purposes of use of the resource, because the regulation involved only irrigation of fields, whereas water for domestic consumption was freely available. Current conflicts over water, however, are more complex and involve different uses – from supplying drinking water, to serving as sewage disposal facilities, to cooling off reactors – as well as conflicts among the diverse users. Chapters 5 and 6 focus on providing guidelines for setting up institutions that would be able to meet such objectives.

5 Transnational institutions for transboundary ecosystem management: defining the tasks and the constraints

The tasks of transnational ecosystem institutions

Chapter 2 concluded with the suggestion that thought should be given to the appropriate international norms that would provide proper incentives for states sharing transboundary natural resources to endogenously establish forms of regional cooperation. As demands for such resources grow and diversify and supplies become increasingly depleted, more and more states may be unable to afford unilateral action and may, therefore, seek cooperation. A body of norms designed to lower the costs involved in such cooperation will hasten its emergence. We began the search for such norms by first noting in chapter 3 that states often fail to cooperate due to domestic conflicts of interests and collusion amongst smaller but politically dominant domestic interest groups. Identifying small groups within states as the prime producers of capture and externalities, chapter 4 offered a remedy in the form of transnational institutions, which could lower the costs of regional cooperation by diminishing the opportunities for small-group domination. Transnational institutions must be tailored to the specific matters they aim to address. Continuing in our quest, this chapter is devoted to an examination of the challenges involved in ecosystem management.

Nowhere is the need to set up transnational institutions more acute than in the sphere of ecosystem management. Management of ecosystems consists of a constant, almost daily balancing of a myriad of demands on a relatively fragile and scarce shared resource. Hence, when the clash between conflicting demands intensifies, the ailments of the Westphalian state system become strongly apparent. What are, in fact, the ailments of that system in the sphere of ecosystem management? How can we contain interest groups' capture? To respond to these questions, this part examines the particular tasks and constraints of ecosystem management. The second part then inquires as to what extent markets can provide satisfactory responses to the tasks and constraints identified in the first part.

The burgeoning literature on environmental and natural resources law has tended to center on such concepts as "biodiversity," "sustainable development," "the precautionary principle," and "inter-generational equity" as principles for sound "ecosystem management." The rhetoric of human rights has also been invoked in this context. This part attempts to flesh out the meanings of those principles as the principles that should guide the shared management of transboundary ecosystems. In order to provide a coherent overview without too much eco-jargon, I focus on three questions: What are the purposes of shared management of transboundary ecosystems? Who are the beneficiaries of such an enterprise? And third, what normative constraints exist in the management of the diverse interests involved? The responses to these questions lay the groundwork for analyzing the "how to" questions to be dealt with in chapter 6.

The subject matter to be managed: natural resources, conflicting claims, and risks

The immediate task of shared ecosystem management is, first and foremost, allocating the shared resource or resources among the diverse uses and users. But ecosystem management is often much more than that. Due to the often versatile character of an ecosystem and its diverse uses, questions of allocation usually entail making value judgments among competing demands. Furthermore, ecosystem management also requires management of risks that are generated by the high level of uncertainty surrounding management decisions. Hence, our discussion of management tasks reflects these three distinct matters: resource management, demand management, and risk management.

Natural resources management

Frequently we speak of several resources that are interlinked in a wide range of ways: sub-basins interact within a larger basin; air, soil, and water are interdependent. The quality of one resource influences that of the others; a problem in one component – for example, deforestation that creates soil erosion – leads to problems in related resources – loss of arable lands and flooding.[1] Use of water for drinking or irrigation reduces the water supply for maintenance of estuarine ecosystems. This interdependency between water, air, and soil is captured by the term "ecosystem" and by the call for "ecosystem management."[2] As Jutta Brunnée and Stephen J. Toope suggest, "An 'ecosystem approach' requires consideration of whole systems rather than individual components. Living species and their physical environments must be recognized as interconnected, and the focus must be on the interaction between different sub-systems and their responses to stress resulting from human activity."[3]

It may seem awkward to adopt such an inclusive approach, which may increase the number of participants in joint management processes and thereby render cooperation more cumbersome. However, although the inclusive approach does, indeed, portend this risk, on the other hand, it also carries the promise of better management opportunities for internalizing the entirety of the consequences of the policies pursued. Collective decision making concerning the complex interrelationships among the various related resources, such as fresh water, air, and soil, as well as concerning the diverse activities and demands with regard to these resources is essential for eliminating the import and export of externalities among members and non-members of the transnational institution. Moreover, the inclusion of control over several resources under the aegis of a single institution often increases the incentive to cooperate. For example, a slanted upstream–downstream relationship with respect to a shared river – the classic example of intransigence – will be rectified if in addition to management of the shared river, the

---

[1] On the link between deforestation, soil erosion and fresh water, see chapter 1, notes 27–9, and accompanying text.

[2] George Francis, "Ecosystem Management" (1993) 33 Nat. Res. J. 316 (describing the emergence of "ecosystem" rhetoric in the management of the Canada–US transboundary region during the 1970s and 1980s, as reflected in the 1978 Great Lakes Water Quality Agreement). On the development of the "Ecosystem approach" since the 1972 Stockholm Conference on the Human Environment see Ludwik A. Teclaff, "Evolution of the River Basin Concept in National and International Law" (1996) 36 Nat. Res. J. 359, 378–81.

[3] Jutta Brunnée and Stephen J. Toope, "Environmental Security and Freshwater Resources: A Case for International Ecosystem Law" (1994) 5 Yb. Int'l Envt'l L. 41, 55. The 1992 Helsinki Convention on the Protection and Use of Transboundary Watercourses and International Lakes (reprinted in [1992] 31 ILM 1312) adopts a broad approach recognizing the link between fresh water and the ecosystem (see the definition of "transboundary impact" in Article 1[2]).

shared institution also encompasses the management of transboundary air pollution drifting in the upstream direction.[4]

The management of shared ecosystems involves a variety of decisions: on the allocation of quantities of the given resource or resources to different consumers; on the quantities and types of pollutants and/or pesticides that industries or farmers may discharge directly or indirectly into the ecosystem; on the establishment of rehabilitation projects, such as waste-water treatment facilities, to protect the affected resource(s) and increase the yield. Shared ecosystems are highly idiosyncratic. Their efficient and equitable management requires intensive investment of resources in long-term planning of both the physical infrastructure and the institutional mechanisms for decision making regarding allocation and pricing, compliance monitoring, enforcement, dispute settlement, and crisis management. This section analyzes and elaborates on the challenges involved in ecosystem management.

Demand management

Several natural resources – including clean air, fresh water, arable lands, and rainforests – are vital to human subsistence. Unlike minerals and other shared resources of strictly economic value, these resources are responsible for a wide array of natural and human processes and have diverse uses, as captured by the *Koran* verse "We made from water every living thing."[5] In contrast to the uses of such natural resources as gas and oil, only a few of the uses of these vital resources can be easily translated into economic value.[6] Beyond subsistence, these resources also provide a plethora of uses, including commercial uses in agriculture, aquaculture, industry, power generation, tourism, and recreation.

---

[4] This is the situation, for example, in the Israeli–Palestinian context: the Palestinians are upstream to a shared aquifer, but the winds blow from Israeli territory to the Palestinian side.

[5] *The Koran*, XCIX, verse 30.

[6] On the management of shared gas and oil resources, see, for example, Rodman R. Bundy, "Natural Resource Development (Oil and Gas) and Boundary Disputes" in Gerald H. Blake *et al.* (eds.), *The Peaceful Management of Transboundary Resources* (London, Graham & Trotman, 1995), p. 23 (maintaining that the law on the exploitation of oil and gas reserves is still based largely on the law of capture, but arguing for a right of consultation and notification); Charles Robson, "Transboundary Petroleum Reservoirs: Legal Issues and Solutions" in *The Peaceful Management of Transboundary Resources*, p. 3 (explaining the fluid nature of oil and gas, the failure of the law of capture, and the solution of field unitization by agreement). See generally Hazel Fox, *Joint Development of Offshore Oil and Gas* (2 vols., London British Institute of International and Comparative Law, 1989).

The diversity of interests in these natural resources and, especially, in fresh water has shaped the wide range of possibilities open to regulators for allocating and managing the resources. The regulation of water or air is different from the regulation of whaling and straddling stocks, of nuclear reactors, and of navigation on international rivers. The former type of regulation is all-encompassing, requiring that a proper balance be struck among the conflicting demands of individuals, groups, and corporate actors and involving diverse concerns, from human subsistence, to economic welfare and cultural needs, to recreational activities. Regulation is constrained by social and cultural factors. In some Muslim countries, for example, religious edicts dictate that water be freely accessible to all, thereby depriving regulators of the possibility of reducing demands through pricing mechanisms. Hence, the traditional and religious objections to full pricing of water prevent the governments of Egypt, Jordan, and Syria from curtailing the rising demands on Allah's gift.[7]

Prominent among the unique challenges faced by regulators of demand and supply management is contending with the policy of ensuring "food security" taken by certain countries. Many governments, particularly in developing countries with unstable water supplies, tend to maintain a strategic interest in food security. They seek to reduce their dependency on foreign supplies. Strong domestic agricultural lobbies often press for adoption of policies aimed at protecting their produce. As a consequence, governments strive to ensure sufficient water supplies for providing fresh produce and milk, which might not be easily obtainable through international trade. The very same agricultural lobbies are responsible for protectionism in international trade norms related to trade in agricultural products. When imports are not a reliable source of basic foodstuffs, a wider margin of food security is necessary, at the cost of raising the demand for water for irrigating crops. A policy of food security in arid and semi-arid countries is wasteful, compared to a

---

[7] Abdul-Karim Sadik and Shawki Barghouti, "The Water Problems of the Arab World: Management of Scarce Resources" in Peter Rogers and Peter Lydon (eds.), *Water in the Arab World* (Cambridge, MA, Cambridge University Press, 1994), pp. 1, 25. See also Chibli Mallat, "Law and the Nile River: Emerging International Rules and the Shari'a" in P. P. Howell and J. A. Allen (eds.), *The Nile: Sharing a Scarce Resource* (Cambridge, Cambridge University Press, 1994) pp. 365, 372–8; Chibli Mallat, "The Quest for Water Use Principles: Reflections on the *Shari'a* and Custom in the Middle East" in J. A. Allen and Chibli Mallat (eds.), *Water in the Middle East: Legal, Political, and Commercial Implications* (London, I. B. Tauris Publishers, 1995), p. 127; A. M. A. Maktari, *Water Rights and Irrigation Practices in Lahj* (Cambridge, Cambridge University Press, 1971).

policy of enhanced trade. But uncertainties with respect to reliance on foreign sources lead many governments to pursue that inefficient policy.

A further concern for governments in many developing countries is the impact that restricted access to water has on peasants' livelihoods. For peasants in these countries, water means employment and maintenance of their livelihoods in their remote villages. Insufficient or costly water supply means dislocation; and governments are concerned with demographic changes and heightened social pressures as cities become crowded beyond capacity. Such considerations may weigh against privatizing resource management. Thus, for example, the establishment in Chile in 1981 of a system of private transactions in water entitlements effectively precluded the poor peasants from this market.[8]

To obtain an optimal and sustainable utilization of vital resources and meet the diversified and constantly increasing demands on these resources, decision-makers must collectively and continuously juggle the manifold and conflicting demands on and supplies of the resources in their ecosystems. Management of ecosystems is, to a large extent, a matter of redistribution of a natural resource, given certain physical, economic, environmental, social, and cultural constraints. It is also a matter of setting priorities among different uses and different users and of designing the optimal structures for making the most of existing supplies.

Often the management of ecosystems entails much more than balancing supplies against demands. At times, ecosystem management requires weighing additional considerations, including even cultural and religious factors when a specific natural resource is a religious symbol for a specific group or when the resource is a necessary element in the preservation of a group's culture.[9] Thus, for example, the preservation of the Saami minority's practice of reindeer husbandry, an essential

---

[8] Carl J. Bauer, "Bringing Water Markets Down to Earth: The Political Economy of Water Rights in Chile, 1976–95" (1997) 25 *World Development* 639, describes Chile's 1981 Water Code as a reflection of a conservative backlash against the agrarian reforms of the 1960s. This Code allowed private transactions in water and a weak regulatory system. On the impact of such concerns on the possibilities of transnational trade in resources see notes 91–4 and accompanying text.

[9] United States federal courts dismissed suits brought by native American tribes against water projects that inundated sacred sites and cemeteries; see, e.g., *Sequoyah* v. *Tennessee Valley Authority*, 620 F.2d 1159 (6th Cir. 1980); *Badoni* v. *Higginson*, 638 F.2d 172 (10th Cir. 1980). In 1997, a Japanese district court invalidated a decision to construct a dam that would have impinged on cultural interests of the Ainu minority: Yuji Iwasawa, *International Law, Human Rights, and Japanese Law* (Oxford, Clarendon Press, 1998), p. 52. See also S. James Anaya, "Maya Aboriginal Land and Resource Rights and the Conflict over Logging in Southern Belize" (1998) 1 Yale HR & Dev. LJ. 17.

element of Sami culture, requires the conservation of forests in Finland.[10] Ecosystem management may thus require attention also to the issue of group representation, especially the representation of indigenous peoples, in decision-making processes.[11]

Among the natural resources, fresh water particularly has nurtured numerous cultures and traditions that have developed over time, reflecting the scarcity of this resource to many communities since ancient times. Because water supplies often fluctuate, especially in semi-arid regions, their availability – in the forms of rain falling from the skies, springs gushing from desert land, or mighty rivers flooding the land – has been closely associated with divine blessing just as their deprivation has been deemed a divine curse. Water is the symbol of life, just as the desert is of death, and is associated with divine ethical judgment.[12] This significance is reflected in the role water or, rather, rivers play in almost all religions, in the tales handed down through the generations, and in the classic literature of all cultures, as well as in the grand schemes of leaders of developing countries throughout the latter half of the twentieth century to harness rivers.[13] This explains the otherwise paradoxical phenomenon that the less water is available, the more it is considered a free communal resource, a divine gift, which, therefore, cannot be subject to personal ownership and must be shared so that it satisfies all living creatures.

---

[10] See the view of the Human Rights Committee under the Optional Protocol of the 1966 International Covenant on Civil and Political Rights (ICCPR) (finding that reindeer husbandry is an essential element of Sami culture and, as such, protected under Article 27 of the Covenant): *Länsman* et al. v. *Finland*, Communication No. 511/1992, UN Doc. CCPR/C/52/D/511/1992 (1994).

[11] See the Draft United Nations Declaration on the Rights of Indigenous Peoples, adopted by the Sub-Commission on Prevention of Discrimination and Protection of Minorities, UN Commission on Human Rights, on 26 August 1994, (reprinted in [1995] 34 ILM 541). The Declaration recognizes the value of water resources to indigenous peoples' social structure, culture, and tradition (see the Preamble). The draft Declaration sets out to ensure, *inter alia*, indigenous peoples' right to maintain and strengthen their relationship with their land, territories, waters, and other resources (Article 25), to own and to manage these resources (Article 26), and their right to participate in decisions affecting these resources. See also Article 22 of the Declaration of the UN Conference on Environment and Development (the "Rio Declaration"), Doc A/CONF 151/5/Rev.1 (1992). See also Benedict Kingsbury, "Claims by Non-State Groups in International Law" (1992) 25 Cornell Int'l L. J. 481. See also discussion in chapter 7, notes 141–6 and accompanying text.

[12] On water in the wilderness as a motif of fertility, see William Henry Propp, "Water in the Wilderness: The Mythological Background of a Biblical Motif" (Ph.D., Harvard University, 1985).

[13] On these schemes, see chapter 1, notes 31–3 and accompanying text.

Risk management

Ecosystem management is fraught with uncertainties and, therefore, with risks. This means that decision making must constantly seek new information and analyses on the possible impact of certain uses and practices on the resource or resources in question. These risks pertain to the impact of diverse activities – use of pesticides and fertilizers in agriculture, different uses of water, water reuse after treatment, water installations (such as dam diversion or irrigation systems), different land uses – on the survival of the ecosystem, the availability of water quantity and quality, and consequential health effects. Some risks are beyond human control, such as those imposed by natural disasters. Other risks are beyond the control of those managing the specific ecosystem, such as the risks involved in climate changes due to global warming. But many other risks may be addressed by the institution in charge of managing the shared ecosystem. Some of these manageable risks can be reduced by, for example, investing in monitoring the compliance of users or providing incentives for users to employ environment-friendly methods. But often such decisions are hampered by uncertainties as to the impacts of alternative uses or the certainty that alternative uses pose alternative risks.

Many known risks can be eliminated only at the cost of increasing the potential of other risks. As a result, decisions often involve weighing the tradeoffs between risks. Take, for example, the ubiquitous question of whether or not to chlorinate drinking water. Studies have indicated that chlorination increases the likelihood of cancer. But reducing the risk of cancer (by ceasing chlorination) will increase the likelihood of microbial diseases.[14] Often, such decisions entail tradeoffs between specific groups in society. If water chlorination is ceased, adults will be at lower risk, while children and the elderly will be more highly exposed. When, for example, the US Environment Protection Agency banned the use of ethylene dibromide (EDB) to prevent the development of molds on grain and other foods, the growers used other fungicides which exposed the workers who applied them to greater hazard than presented by EDB.[15]

[14] Susan W. Putnam and Jonathan Baert Wiener, "Seeking Safe Drinking Water" in John D. Graham and Jonathan Baert Wiener (eds.), *Risk Versus Risk: Tradeoffs in Protecting Health and the Environment* (Cambridge, MA, Harvard University Press, 1995). Frank B. Cross, "Paradoxical Perils of the Precautionary Principle" (1996) 53 Wash. & Lee L. Rev., 851, 883, provides evidence that when Peru responded to the chlorine scare by halting the chlorination of the water supply, thousands died of cholera.

[15] See Cross, "Paradoxical Perils," note 14, at pp. 875–6.

Decisions also could involve tradeoffs between countries, as, for example, would be the case where political boundaries separate growers and consumers.

Science cannot eliminate the uncertainties surrounding the potential risks of certain uses of a shared ecosystem.[16] In some cases, experiments can establish relationships of cause and effect. Often, however, it is unclear how to evaluate the laboratory findings and what measures should be adopted based on that knowledge. Not only does science fail to resolve many uncertainties, it in fact frequently creates new ones, as new findings question the safety of previous policies, or as new technologies generate new risks. Thus, science does not release decision-makers – whether legislators, regulators, courts, or individuals – from applying discretion in the adoption of policies. Management of shared resources, therefore, involves not only allocation of shares in the scarce resource among its several users: it also entails allocation of the known risks posed by that allocation of shares, as well as allocation of the unknown or insufficiently researched risks of the approved uses and practices.

Sometimes the good intention to avoid one risk leads unwittingly to an even graver risk. In Bangladesh and in the neighboring Indian state of West Bengal, tens of thousands of villagers are slowly being killed by the water they drink because it contains high levels of natural cancer-causing arsenic. The villagers haul the water by hand-operated pumps, distributed as part of a joint effort by UNICEF and the Bangladeshi and the Indian governments to end the villagers' reliance on polluted surface water. The surface water used previously by the villagers caused widespread and deadly diarrhea and cholera, claiming the lives of hundreds of thousands each year. Instead, the pumps draw water from a shallow aquifer, which has turned out to be naturally contaminated. Neither UNICEF nor either of the governments ever tested the aquifer and were slow to react to scientific warnings already sounded in 1988. Still, villagers are slow to demand change: most of them do not know what they are drinking and attribute their diseases to supernatural causes.[17]

In environmental law – both domestic and international – the function of risk management is captured in the so-called "precautionary principle." Under this principle, "where there are threats of serious or

---

[16] On the uncertainties involved in water management, see Larry Canter, Konrad Ott, and Donald A. Brown, "Protection of Marine and Freshwater Resources" in John Lemons and Donald A. Brown (eds.), *Sustainable Development: Science, Ethics, and Public Policy* (Dordrecht, Kluwer Academic Publishers, 1995).

[17] "Death By Arsenic," *New York Times*, 10 Nov. 1998, at A1.

irreversible damage, lack of full scientific certainty shall not be used as reason for postponing cost-effective measures to prevent environmental degradation."[18] This principle appeals to the human instinct to follow such rules of thumb as "Better safe than sorry" and "Do no harm."[19] It calls upon decision-makers to proscribe uses that could harm the environment or deplete resources, until the harmless effects of such uses are adequately proven.

The precautionary principle underscores the important role of risk management as integral in managing the environment and other shared resources. But this principle is, at best, irrelevant in terms of determining just how, precisely, risks should be managed. It fails to define the level of uncertainty that triggers the principle into operation; who bears the onus of proving such uncertainty; and which principles should inform decision-makers in their cost/benefit analyses. Moreover, beyond this vagueness, reliance on the precautionary principle may prove counterproductive to proper resource management. As unfortunately is too often the case, this principle is invoked to uphold the status quo. Under this rendition, the principle contributes negatively to sound resource management, which must constantly consider the tradeoffs entailed in any potential use. While it might not be sufficiently clear what the adverse consequences of a certain practice may be, it might be equally unclear about the adverse consequences of *prohibiting* that same practice or of resorting to an alternative one. As demonstrated by the above examples of water chlorination and EDB application to grains, managers should almost always weigh the adverse consequences of permitting a certain usage (chlorination, EDB application) against the adverse consequences of proscribing that usage. It is often rather uncertain whether "Better safe than sorry" is preferable to taking a calculated risk in anticipation of an unfolding crisis or to improving the public's general health and living conditions. Uncertainties exist either way, and

---

[18] Article 15 of the Rio Declaration (note 11). See also Article 5(a) of the 1992 Helsinki Convention on the Protection and Use of Transboundary Watercourses and International Lakes, reprinted in (1992) 31 ILM 1312: "action to avoid the potential transboundary impact of the release of hazardous substances shall not be postponed on the grounds that scientific research has not fully proved a causal link between those substances, on the one hand, and the potential transboundary effect, on the other hand." On the origins of the principle and its standing in contemporary international law, see, e.g., David Freestone and Ellen Hey (eds.), *The Precautionary Principle and International Law: The Challenge of Implementation* (The Hague, Kluwer Law International, 1996).

[19] See Howard Margolis, *Dealing with Risk* (Chicago, IL, University of Chicago Press, 1996) (people tend to be prudent, follow rules like "Do no harm" or "Better safe than sorry").

hesitant decision-makers may err on the side of inertia only later to regret it.[20]

When new uses of a resource arise for consideration, the precautionary principle serves as an obstacle to a potentially beneficial practice. Take, for example, the question of whether to allow farmers to seed genetically modified grains. Such grains are more resilient than others against pests; they do not require the application of pesticides and, hence, reduce pollution to the watercourse. Losses due to weeds and pests are reduced significantly, thereby increasing the yield substantially and, at the same time, either increasing productivity or allowing managers to redirect part of the designated water to other beneficial uses. However, the effects of these grains on human health have yet to be established. The "Better safe than sorry" rule of thumb would, thus, ban use of genetically modified grains. But this ban might then be responsible for insufficient food and water supplies. Faced with this dilemma, the precautionary principle is invoked by the two opposing sides, each with its own version of the better way to manage the ecosystem: while the group opposing genetic modifications calls attention to its potential adverse health effects, the other group warns against the continued waste and degradation associated with current grains and practices.[21] It is highly questionable whether the status quo should be venerated, especially since there is no status quo in an ever-changing environment. Population growth, urbanization, occasional disasters, recurring droughts – all constantly push and pull at the existing arrangements and call for their periodic reassessment. In a constantly metamorphosing ecosystem, inaction has no inherent superiority over action. In this sense, the precautionary principle, whose baseline is the current situation, does not call into question the status quo and hence is ill-equipped to assess the demands to modify that status quo. The precautionary principle conveys the potentially erroneous message that present uses are *a priori* safe.

The enthusiastic and unequivocal embracing of the precautionary principle by environmentalists and scholars of domestic and

---

[20] See Cross, "Paradoxical Perils," note 14 (questioning the precautionary principle by showing many instances where environmental and public health regulations, based on this principle, frequently produce health or other environmental harms); Daniel Bodansky, "Scientific Uncertainty and the Precautionary Principle" (Sept. 1991) 33 *Environment* 4 (critical of the principle due to its indeterminacy as to when it must be applied, what it requires be done and at what cost).

[21] See, e.g., BBC Special Report, "Friend or Foe?" 17 Mar. 1999, at http://news.bbc.co.uk.

international environmental law[22] highlights the dangers of risk assessment. These dangers include the inherent tendency of the general public to err in assessing risks, the inclination of officials to "do no harm," the potential manipulation of people's perceptions of risks by interest groups, and – in the international context – the different attitudes of different societies towards risk aversion in general and specific risks in particular.

Human error in assessing risks results from the observed tendency to base the assessment of probabilities on biased methods ("heuristics" in the current jargon). Human intuition may economize one's resources in predicting future developments, but as Daniel Kahneman and Amos Tversky demonstrated in a series of experiments they conducted, intuition often leads us to use patently wrong methods that result in wrong assessments.[23] Human judgment is, for example, insensitive to the size of a sample. People disregard the fact that the larger the sample (say, a sample of balls from a bottle containing red and blue balls), the more accurate its representation of the entire population (all the balls inside the bottle).[24] By adhering to "the law of small numbers," people tend to formulate generalizations based on a rather limited number of events, just as they expect the tossed coin to land on tails after a very short sequence where it landed on heads.[25] Events that are fresh in their memories, such as recent major catastrophes, or which have made a strong impact due to personal involvement, such as a car accident, increase the subjective probability of their recurrence.[26] People's subjective assessments are influenced by a variety of other conditioning factors, such as the favorable or unfavorable description of the activity or the actor (say the future value of a company),[27] or by an arbitrarily chosen event that they use as the starting point for their assessment.[28] The status quo often serves

---

[22] See, e.g., David Freestone and Ellen Hey, "Origins and Development of the Precautionary Principle" in Freestone and Hey, *Precautionary Principle,* note 18, at pp. 3, 4 ("The challenge facing the international community is how to attain truly precautionary environmental policies"); Brunnée and Toope, "Environmental Security," note 3, at 68 (referring to this principle as "the single most important underpinning of any regime intended to promote ecological balance and ecosystem integrity"). For a refined and sensitive treatment of this principle, see André Nollkaemper, "'What You Risk Is What You Value', and Other Dilemmas Encountered in the Legal Assaults on Risks" in Freestone and Hey, *The Precautionary Principle,* note 18, at pp. 73–94.

[23] See Amos Tversky and Daniel Kahneman, "Judgment under Uncertainty: Heuristics and Biases" in Daniel Kahneman, Paul Slovic, and Amos Tversky (eds.), *Judgment Under Uncertainty* (Cambridge, Cambridge University Press, 1982), pp. 3–31.

[24] *Ibid.,* at pp. 6–7.   [25] *Ibid.,* at pp. 7–8.   [26] *Ibid.,* at p. 11.

[27] *Ibid.,* at p. 8.   [28] *Ibid.,* at pp. 14–18.

as just such a starting point. More specifically, in the context of health risks, experiments have established that people tend to underestimate certain health risks while overestimating other health risks. People tend to adopt the worst-case scenario. They respond to exaggerated small hazards with excessive intransigence and take precautions accordingly.[29]

This "bounded rationality" falls prey to manipulation by different interest groups.[30] These groups employ two strategies in promoting their interests. They either supply bureaucrats and politicians with partial or skewed "scientific findings" to support their demands or – if their pressure is resisted by the latter – they intimidate the public with doomsday predictions against a suggested policy. The first strategy is usually resorted to when officials seek ways to promote sectarian interests, but have few opportunities to offer outright benefits such as import quotas or duties. "Scientific" grounds for refusing to grant import permits, for example, become the basis for the seemingly reasonable exercise of governmental discretion which in fact protects domestic industries. Too often, both in the domestic[31] and the international[32] arenas, "science" is invoked to mask partisan politics. Alleged scientific findings, sometimes devoid of any real scientific value, are often used to provide the government with ostensible "objective" grounds to grant concessions to interest groups.

On the other hand, when bureaucrats are less accommodating, pseudo-science is used to turn public opinion against them. The scares

---

[29] W. Kip Viscusi, *Rational Risk Policy: The 1996 Arne Ryde Memorial Lectures* (Oxford, Clarendon Press, 1998), pp. 5–25. See also Viscusi *Fatal Tradeoffs: Public and Private Responsibilities for Risk* (New York, Oxford University Press, 1992), pp. 111–29.

[30] On the manipulation of consumers by producers, see Jon D. Hanson and Douglas A. Kysar, "Taking Behavioralism Seriously: Some Evidence of Market Manipulation" (1999) 112 Harv. L. Rev. 1422.

[31] Robin Shifrin, "Not by Risk Alone: Reforming EPA Research Priorities" (1992) 102 Yale L. J. 547, 558–9 ("Risk based budgeting has the potential for adding the legitimacy of science to executive politics."). See also Wendy E. Wagner, "The Science Charade in Toxic Risk Regulation" (1995) 95 Colum. L. Rev. 1617; Donald T. Hornstein, "Reclaiming Environmental Law: A Normative Critique of Comparative Risk Analysis" (1992) 92 Colum. L. Rev. 562.

[32] There are numerous examples of such practices in the context of non-tariff barriers to international trade. See Michael J. Trebilcock and Robert Howse, *The Regulation of International Trade* (2nd edn, London, Routledge, 1999), chapter 6; Julie A. Soloway, "Environmental Trade Barriers Under NAFTA: The MMT Fuel Additives Controversy" (1999) 8 Minn. J. Global Trade 55; Vern R. Walker, "Keeping the WTO from Becoming the 'World Trans-science Organization': Scientific Uncertainty, Science Policy, and Factfinding in the Growth Hormone Dispute" (1998) 31 Cornell Int'l L. J. 251; David A. Wirth, "The Role of Science in the Uruguay Round and NAFTA Trade Disciplines" (1994) 27 Cornell Int'l L. J. 817, 834–5.

concerning genetically modified food, the injection of hormones into cattle, the use of Alar as a growth hormone for apples, and the use of MMT as an additive in gasoline are all examples of potential health risks that, once brought to the public's attention, created an uproar that led to immediate reaction on the part of the authorities before the merits of these claims could be examined. In fact, the risks entailed in Alar eventually were proven to be "somewhere between miniscule and nonexistent";[33] evidence was brought to show that the risks of the MMT additive were potentially lower than those of alternative additives;[34] the Joint Expert Committee on Food Additives of WHO and the Food and Agriculture Organization approved the safety of certain growth hormones for cattle; and the Royal Society, Britain's foremost scientific institution, criticized as flawed a study that purported to show that genetically modified crops are unsafe.[35] Despite these counter assessments of the risks involved, public suspicion has persisted.

One way or another, science is used and abused by better-organized groups, usually the smaller, well-off groups, to shift burdens onto the lesser-endowed groups. Not only do the better-off groups tend to have lower transaction costs; they tend also to have a different, higher aversion to health risks, due to the decreasing marginal utility of their income that allows them to spend more on health and safety and, therefore, a greater interest in averting or externalizing the risks.[36] They thus have better prospects of influencing decision-makers to reallocate public funds for environmental conservation and rehabilitation projects, to reallocate public lands to polluting or potentially risky facilities, and to shift environmental and health risks from consumers to workers in industry or agriculture. Furthermore, the voice of future generations is repressed in the political process.

All these diverse interests work to pull government officials towards compliance. Public measures and new regulations are prescribed in

---

[33] See Cross, "Paradoxical Perils," note 14, at p. 886; Eliot Marshall, "A Is for Apple, Alar and . . . Alarmist?" (4 Oct. 1991) 254 *Science* 20, 20. The editorial was critical: "a clearly dubious report about possible carcinogenicity by a special interest group was hyped by a news organization without the most simple checks on its reliability or documentation."

[34] Soloway, "Trade Barriers," note 32, at p. 93 (suggesting that alternative additives, advocated by environmentalists, could be even more harmful).

[35] See the BBC News Service, *GM Food Report Condemned as Flawed*, 18 May 1999, at http://news2.thdo.bbc.co.uk.

[36] See Frank B. Cross, "When Environmental Regulations Kill" (1995) 22 Ecology L. Q. 729, 734–5 (setting forth the theory that greater wealth tends to lead to greater concern for health protection).

response to public outcries. "Responding to irrational public pressures may be 'rational' behavior but bad policy."[37] The cautious tendency of officials to be "Better safe than sorry" or otherwise err in their management of risk is not only the result of public pressure or interest group capture or of a combined, so-called "Baptist–bootlegger" coalition.[38] It is also a consequence of the officials' own bounded rationality, which is heightened by their aversion to be later identified as responsible for a risk-creating policy. Thus, as Kip Viscusi observed, "One FDA official remarked that no one was going to blame him for slow approval of a beneficial drug, but he would be blamed for approving the next thalidomide."[39] Sometimes, ostensibly uncertain health and other effects disguise hidden moral and policy agendas. The Japanese advisory council to the Health and Welfare Minister, for example, recommended allowing the use of birth control pills only in June 1999, ending a review process of more than nine years, amidst allegations that the long-deferred approval reflected concerns with Japan's declining birth rate, which may even have been gender-biased, in light of the swift approval of the Viagra pill one month before.[40]

Finally, the divergences involved in the existence of several regulatory agencies provide regulators with incentives and constraints that distort their choices and lead them to disregard or underestimate certain risks. The limits of their regulatory authority require them to adopt "tunnel vision" that focuses only on a fragment of the potential risks[41] and, at the same time, allows them to externalize the risks of their own regulations onto other agencies.[42] Serving bureaucrats also have an inherent and deeply rooted preference for maintaining the bureaucratic status quo, and they are averse to any rearrangement of agency powers in ways that could reduce their authority or budgets. They are equally averse

---

[37] See Christine Jolls, Cass R. Sunstein and Richard Thaler, "A Behavioral Approach to Law and Economics" (1998) 50 Stan. L. Rev. 1471, 1520–3 (discussing the success of the Superfund legislation despite its uncertain empirical support).

[38] Soloway, "Trade Barriers," note 32, at p. 59 (discussing whether the Canadian Act that prohibited MMT resulted from the combined pressure of environmentalists and local industry).

[39] Viscusi, *Fatal Tradeoffs*, note 29, at p. 273 (1992).

[40] "Japan, Never on the Pill, Seems Ready to Try It," *New York Times*, 3 June 1999, at A11.

[41] Stephen Breyer, *Breaking the Vicious Circle: Toward Effective Risk Regulation* (Cambridge, MA, Harvard University Press, 1993), pp. 10–28; John S. Applegate, "A Beginning and not an End in Itself: The Role of Risk Assessment in Environmental Decision-Making" (1995) 63 U. Cin. L. Rev. 1643, 1652.

[42] Jonathan Baert Wiener and John D. Graham, "Resolving Risk Tradeoffs" in Graham and Wiener, *Risk versus Risk*, note 14, at p. 226.

to spending resources on research that could reduce uncertainties that serve their interests.[43] For one reason or another, or any combination thereof, agencies tend to prefer a status-quo-oriented, "Better safe than sorry" policy in managing risks.

The difficulties mentioned above multiply in the transnational context. Different perceptions of risks and different aversions to risks can result both from economic and cultural differences. More importantly, the incentive to impose risks on the neighboring community, to whom there is no accountability, is great in interstate settings.

The beneficiaries of transnational ecosystem institutions

The beneficiaries of transboundary resource management are the individuals who rely on the resource for whatever purposes. But their interests frequently clash. Hence, transnational institutions must prioritize the different and often conflicting claims.

Out of these individuals, special attention should be given to two groups that are disadvantaged in many societies: future generations and members of minority groups.[44] Equity requires due regard to the demands of future generations as well as to the often unique interest of minority groups – especially indigenous peoples – in the preservation of natural resources. Efficiency also supports this requirement. As discussed in chapter 1, the short-term policies of many governments and the sometimes gross human rights violations associated with such policies in relation to the management of natural resources place in doubt the willingness and ability of states to commit themselves to long-term cooperation. When riparians resort to short-term policies, policies that reflect internal political instability or create such instability in their infringement of human rights, the incentive of neighboring riparians to cooperate diminishes.

The demands of future generations pose unique questions. The principle that future generations are entitled to equal concern and respect in making decisions concerning the ecosystem is widely accepted.[45] It is recognized as a legal obligation in both national legal systems[46] and

---

[43] Shifrin, "Not by Risk," note 31, at p. 565.
[44] On the unique interests and rights of minority groups in the context of ecosystem management, see notes 9–11 and accompanying text.
[45] John Rawls, *A Theory of Justice* (Oxford, Clarendon Press, 1972), p. 137.
[46] For reference to this principle by national systems, see the 1997 South African Constitution, section 24(b) ("Everyone has the right – . . . [b] to have the environment

international law.[47] The International Court of Justice recognized that "generations unborn" are among those whose "quality of life and the very health depend on sound environment."[48] In fact, one of the early instances in which reference to the rights of future generations was raised was in relation to a shared river. In 1957, while Egypt and Sudan were negotiating their Nile Waters Agreement, the government of Ethiopia declared that it was reserving its rights for future use of 84 percent of the flow, using the following argument:

Just as in the case of all other natural resources on its territory, Ethiopia has the right and obligation to exploit the water resources of the Empire and indeed, has the responsibility of providing the fullest and most scientific measures for the development and utilisation of the same, for the benefit of present and future generations of its citizens, in pace with and in anticipation of the growth in population and its expanding needs.[49]

Attention to future generations lies at the heart of the principle of sustainability. The management of the environment must be sustainable, because the current generation is the trustee of the natural resources rather than their sole owner. Sustainability does not proscribe efforts to improve current conditions for users. Hence, the term "sustainable development" has gained prominence in various international instruments

protected, for the benefit of present and future generations"), and the Philippine Supreme Court's decision in the case of *Minors Oposa* v. *Secretary of the Department of Environment and Natural Resources*, reprinted in (1994) 30 ILM 173.

[47] Institut de droit international, Resolution on International Responsibility and Liability under International Law for Environmental Damage (Preamble): (1997) 67 Ann. Inst. Dr. Int'l 311; Edith Brown Weiss, *In Fairness to Future Generations: International Law, Common Patrimony and Intergenerational Equity* (Tokyo, Japan, United Nations University, 1989); Alexander Gillespie, *International Environmental Law Policy and Ethics* (Oxford, Clarendon Press, 1997), pp. 107–26; Agora: "What Obligations Does Our Generation Owe to the Next? An Approach to Global Environmental Responsibility" (1990) 84 AJIL 190.

[48] *Legality of the Threat or Use of Nuclear Weapons*, Advisory Opinion, ICJ Reports 1996, p. 226 at 241–2, para. 29. Judge Weeramantary has been the notable proponent on the court of this idea (see his opinions in *Maritime Delimitation in the Area between Greenland and Jan Mayen (Denmark* v. *Norway)*, 1993 ICJ Reports 38, 277 n.1; *Request for an Examination of the Situation in Accordance with Paragraph 63 of the Court's Judgment of 20 December 1974 in the Nuclear Tests (New Zealand* v. *France) Case*, ICJ Reports 1995 288, 341; Advisory Opinion on the Legality of the Threat or Use of Nuclear Weapons, *supra*, and the *Case concerning the Gabcikovo-Nagymaros Project (Hungary/Slovakia)* (1997), ICJ Reports 1997, p. 7, reprinted in (1998) 37 ILM 167.

[49] Aide Memoire of 23 September 1957, circulated to diplomatic missions at Cairo, reprinted in Marjorie M. Whiteman, *Digest of International Law* (15 vols., Washington, US Dept. of State, 1964), vol. III, pp. 1011, 1012.

in the last two decades[50] and has been endorsed by the International Court of Justice.[51] However, this term gives managers only vague guidance regarding how to strike the proper balance between conservation and development.[52]

While the validity of the principles of inter-generational equality or sustainable development is uncontested, the practical meaning of these principles is less certain. First, it is not clear to whom exactly "future generations" refers: Does it include the immediate successive generation, a few generations down the line, or continue *ad infinitum*? Second, what must the present generation actually do?[53] To what extent may the current generation modify the environment it bequeaths to the future generations? Can it offset the deteriorating ecosystem against the improving economic and technological conditions and can it defer the protection of the ecosystem to a later stage, when economic conditions are improved? Who – we, they, or both – should bear the burden created by our predecessors who misused or depleted our resources?[54] Finally, and more concretely, which discount rates should we use in evaluating the costs of investments necessary to prevent future harms?[55]

These are difficult questions that can stir deep philosophical debates.[56] In practice, however, these questions call for striking some sort of balance between the conflicting interests. Different societies will respond differently to many of the questions. Developing and developed societies pursue different sets of priorities. No *a priori* rule can be imposed

[50] World Commission on Environment and Development, *Our Common Future* (1987) ("The Brundtland Report").

[51] *Gabcikovo Nagymaros Case*, note 48, para. 140 ("This need to reconcile economic development with protection of the environment is aptly expressed in the concept of sustainable development.")

[52] A. Dan Tarlock, "Exclusive Sovereignty Versus Sustainable Development of a Shared Resource: The Dilemma of Latin American Rainforest Management" (1997) 32 Tex. Int'l L. J. 37.

[53] See Lothar Guendling, "Our Responsibility to Future Generations" (1990) 84 AJIL 207, 210 ("The major problem is what we have to do today to meet our responsibility to future generations").

[54] One approach is suggested by Brown Weiss, *In Fairness*, note 47, at pp. 37–8 ("each generation has an obligation to future generations to pass on the natural and cultural resources of the planet in no worse condition than received and to provide reasonable access to the legacy of the present generation").

[55] On this problem, which economic theory cannot determine, see Daniel A. Farber and Paul A. Hemmersbaugh, "The Shadow of the Future: Discount Rates, Later Generations, and the Environment" (1993) 46 Vand. L. Rev. 267; Edward R. Morrison, "Judicial Review of Discount Rates Used in Regulatory Cost-Benefit Analysis" (1998) 65 U. Chi. L. Rev. 1333.

[56] Rawls, *A Theory of Justice*, note 45, at pp. 118–42.

on decision-makers. Hence, choosing the responses should be relegated to the participants in the decision-making processes within the relevant institutions.

Normative constraints on transnational ecosystem institutions

Not only nature imposes constraints on transnational ecosystem institutions. Basic normative constraints – resulting from either the national constitutional order of participating states or from international law – must be respected as well. These constraints are imposed on the institutions either directly or indirectly, as a consequence of the normative constraints imposed on the state parties to the institution. Despite the transfer of significant authority to supranational joint management bodies, the participating governments are required by their national legal orders to ensure that such bodies do not infringe their citizens' human rights that are protected by the national constitutional order. This principle has long been acknowledged and developed by national constitutional courts, particularly the German Constitutional Court, in response to domestic challenges to the transfer of powers to institutions of the European Communities.[57] The European Court of Human Rights (ECHR), followed the same logic when it found the state parties to the European Convention on Human Rights responsible for any violations of that convention by the institutions of the European Communities to which they had transferred authority.[58] In its view, "The [ECHR] does not exclude the transfer of competences to international organisations provided that [ECHR] rights continue to be 'secured'. Member States' responsibility therefore continues even after such a transfer."[59] The same principle can be deduced from the international obligation of states, under the general human rights conventions to respect the internationally recognized human rights of all subject to their jurisdiction.[60] Hence, governments

---

[57] See the *Maastricht Treaty 1992 Constitutionality Case*, EuGRZ 1993, 429 (Fed. Const. Ct. FRG), translated in (1994) 33 ILM 388, 420.

[58] *Case of Matthews v. The United Kingdom* (Application no. 24833/94) (18 February, 1999), ECHR 1999-I; 28 Eur. Ct. H. R. 361 (1999).

[59] *Matthews v. The United Kingdom* at para. 32.

[60] Article 2(1) of the 1966 Covenant on Civil and Political Rights; Article 6 of the 1965 Convention on the Elimination of All Forms of Racial Discrimination; Article 2(1) of the 1984 Convention Against Torture and other Cruel, Inhuman or Degrading Treatment or Punishment; Article 1 of the 1950 European Convention on Human Rights and Article 1(1) of the American Convention on Human Rights. See also Thomas Buergenthal, "To Respect and Ensure: State Obligations and Permissible Derogations" in Louis Henkin (ed.), *The International Bill of Rights: The Covenant on Civil*

must ensure that the transnational institution conforms both with national and international norms protecting the human rights of their citizens and of others within their jurisdiction.

Transnational institutions, as subjects of international law,[61] are also constrained directly by basic norms of international law.[62] Besides the rather vague mandate to pursue outcomes that would be optimal and sustainable, the major constraint on the discretion of the transnational decision-makers imposed by international law is the duty to respect and ensure basic human rights.[63] Thus, either through national law or international law, indirectly as a duty borne by participating states or directly imposed on the institution, transnational ecosystem institutions are bound by the obligation to respect and ensure the human rights of those subject to their territorial jurisdiction.[64]

There are only a few references in existing national constitutions and international human rights conventions linking human rights and ecosystem management. The 1989 Convention on the Rights of the Child enumerates the right to "clean drinking-water" among the rights of the child.[65] The 1997 South African Constitution stipulates that "everyone has the right – (a) to an environment that is not harmful to their health or well-being; and (b) to have the environment protected, for the benefit of present and future generations, through reasonable legislative and other measures that – (i) prevent pollution and ecological degradation; (ii) promote conservation; and (iii) secure ecologically sustainable development and use of natural resources while promoting justifiable economic and social development."[66] Another section guarantees the right "to have access to . . . sufficient food and water."[67] Despite the dearth of explicit individual rights, it is widely accepted that such rights derive

---

and Political Rights (New York, Columbia University Press, 1981), pp. 72, 74; Eyal Benvenisti, "The Applicability of Human Rights Conventions to Israel and to the Occupied Territories" (1992) 26 Israel L. Rev. 24, 33–5.

[61] See Reparation for Injuries Suffered in the Service of the United States (Advisory Opinion), 1949 ICJ, 173, 178–9 (April 11); Henry G. Schermers and Niels M. Blokker, International Institutional Law: Unity Within Diversity (3rd edn, The Hague, M. Nijhoff, 1995), pp. 29–45.

[62] Peter H. F. Bekker, The Legal Position of Intergovernmental Organizations (Dordrecht, M. Nijhoff, 1994), pp. 54–61.

[63] For an analysis of these constraints, see chapter 7.

[64] I do not discuss here the obligations of the institution towards affected third parties. On this matter see Moshe Hirsch, The Responsibility of International Organizations Toward Third Parties: Some Basic Principles (Dordrecht, M. Nijhoff, 1995).

[65] Convention on the Rights of the Child, 1989, Article 24(2)(c).

[66] 1997 South African Constitution, section 24.

[67] 1997 South African Constitution, section 27.

from well-recognized, more general provisions.[68] The right to drinking water can be deduced from the right to life and the right to be free from inhumane or degrading treatment,[69] as well as from the right to food.[70] The link between natural resources and the human right to life has been recognized by the International Court of Justice, when it declared that "the environment is not an abstraction but represents the living space, the quality of life and the very health of human beings, including generations unborn."[71] These individual rights are, at the same time, positive duties of the authorities, including the duty to improve current conditions and to provide – whenever possible – an adequate supply of good-quality water and food.[72]

Note, however, that in contrast, the 1997 United Nations Convention on the Law of the Non-Navigational Uses of International Watercourses ("the Watercourses Convention")[73] missed a golden opportunity to make a strong commitment to the human rights aspect. The very most that it did was to recognize the states' duty to give "special regard . . . to the requirements of vital human needs"[74] in the event of a conflict between different uses of an international watercourse. One of the *Statements of Understanding pertaining to certain Articles of the Convention* addresses that

---

[68] Stephen S. McCaffrey, "A Human Right to Water: Domestic and International Implications" (1992) 5 Georgetown Int'l Envt'l L. Rev. 1, 12; Dinah Shelton, "Human Rights, Environmental Rights, and the Right to Environment" (1991) 28 Stan. J. Int'l L. 103.

[69] As recognized, for example, by Articles 6 and 7 of the 1966 International Covenant on Civil and Political Rights. On a similar inference with respect to the right to food, see Philip Alston, "International Law and the Human Right to Food" in Philip Alston and Katarina Tomasevski (eds.), *The Right to Food* (Boston, M. Nijhoff, 1984), pp. 9, 24–5. Alston refers to the General Comments of the Human Rights Committee with regard to Article 6, which urge states to take positive measures to protect the right, including "measures to eliminate malnutrition and epidemics" (Report of the Human Rights Committee, UN Doc. A/37/40 (1982) Annex V, para. 5).

[70] 1966 International Covenant on Economic, Social and Cultural Rights, Article 11. For a thorough legal analysis of this right, see Alston, "Right to Food," note 69, at pp. 9, 29–49.

[71] *Legality of the Threat or Use of Nuclear Weapons*, Advisory Opinion, ICJ Reports 1996, at 241–2, para. 29. See also the *Gabcikovo Nagymaros Case*, note 48.

[72] On the difference between "negative" human rights and "positive" rights (which impose on state parties obligations to provide, *inter alia*, food), see, for example, Godfried van Hoof, "The Legal Nature of Economic, Social and Cultural Rights: A Rebuttal of Some Traditional Views" in Alston and Tomasevski, *The Right to Food*, note 69, at p. 97; Guy S. Goodwin-Gill, "Obligations of Conduct and Result" in Alston and Tomasevski, *The Right to Food* at p. 111; E. W. Vierdag, "The Legal Nature of the Rights Granted by the International Covenant on Economic, Social and Cultural Rights" (1978) 9 Netherlands Yb. Int'l L. 69.

[73] Reprinted in (1997) 36 ILM 700.   [74] Article 10(2).

provision, suggesting that "in determining 'vital human needs', special attention is to be paid to providing sufficient water to sustain human life, including both drinking water and water required for production of food in order to prevent starvation."[75] Although vital human needs become in the Watercourses Convention one of the relevant factors in determining how the transboundary freshwater resource is to be used, such vital needs are not accorded the status of individual entitlements.[76]

The human rights perspective sets three guidelines for decision-makers in structuring and implementing the procedures of the transnational ecosystem institutions. First, the decision-makers are under a clear obligation to provide a safe environment for all individuals who depend on the managed ecosystem and to ensure them the minimum share of good-quality fresh water for a decent human subsistence. States may not agree to join institutions that provide less than this minimum to their citizens. This minimum share must include enough water for domestic uses, for drinking and sanitation, as well as enough water to produce food in those countries that have insufficient resources for importing food.[77] Because this duty extends to the good of both the present and future generations,[78] it prescribes only uses that are sustainable. States also are prohibited from agreeing to impose, via the transnational institutions, unreasonable risks to the lives of their citizens resulting from poor risk management or unequal risk allocation.

The second guideline calls for equal treatment of the individuals dependent on the resource. Transnational institutions may not discriminate between different individuals and communities in providing access to such resources and in the allocation of quantities and risks.[79]

---

[75] Note 73, at p. 719.

[76] For a discussion of the Watercourses Convention, see chapter 7.

[77] In particular, this consideration will require enough water to provide fresh produce and milk, which may not be easily obtained through international trade. On the supremacy of domestic needs, see the Commentary to Article 6 of the International Law Association's Helsinki Rules of 1966 (ILA Report of the Fifty-Second Conference, [1967] 484, 491–2): "If a domestic use is indispensable – since it is in fact the basis of life – it would not have difficulty in prevailing on the merits against other uses in the evaluation of the drainage basin." See also Gunther Handl, "The Principle of 'Equitable Use' as Applied to Internationally Shared Natural Resources: Its Role in Resolving Potential International Disputes over Transfrontier Pollution" (1978) 14 Rev. Belge Dr. Int'l 40, 51–2; *Connecticut v. Massachusetts*, (1931) 282 US 660, 673 ("Drinking and other domestic purposes are the highest uses of water").

[78] On this duty, see notes 45–56 and accompanying text.

[79] On the duty to share water equally for domestic uses in the Israeli–Palestinian context, see Eyal Benvenisti and Haim Gvirtzman, "Harnessing International Law to Determine Israeli–Palestinian Water Rights" (1993) 33 Nat. Res. J. 543, 561; Sharif S.

Decisions on projects that could lead to the displacement of populations – for example, dam construction – must undergo strict and careful scrutiny, give voice to the potentially affected populations during the decision-making process, and provide full compensation, including a resettlement scheme, acceptable to the majority of the displaced people.

Finally, because ecosystems also are often crucial for preserving indigenous societies, their cultural rights, as recognized under international and national law, must be respected.[80] The right of religious or cultural minority groups to the conservation of specific water-related sanctuaries may be founded on their internationally recognized cultural and religious rights.[81] Managers should therefore give ample weight to the demands of indigenous peoples to maintain and strengthen their relationship with their land, territories, waters, and other resources and to their right to participate in decisions affecting these resources.

These human and group rights place major constraints upon states and, hence, upon the transnational institutions' margin of possible outcomes. Any decision with potential impact on the transboundary resource must undergo careful scrutiny and a balancing of the conflicting rights and interests. For example, the damming of rivers or the diversion of flows from one basin to another may increase the availability of water for some people but, at the same time, create adverse environmental and social effects for others. Such cases will necessitate reaching an equitable balance between the interests of the different communities.

Meeting the tasks and the constraints through privatization?

One conclusion from the above analysis is that the tasks that ecosystem management institutions must perform are complex. Their options are constrained by nature, by the interdependency among different uses and users, by uncertainties, and by normative limitations. As we move to examine what type of institutions would best meet these challenges, we first face a fundamental choice between two basic types of institutions: a regulatory command and control regime or a market regime. Note that there is no black and white distinction between the two, because as we

Elmusa, "Dividing the Common Palestinian–Israeli Waters: An International Water Law Approach" (Spring 1993) 22 J. Palestine Studies 3, 57, 68–9.

[80] Article 26 of the 1966 ICCPR.

[81] Article 27 of the 1966 ICCPR. On minorities' dependence on water resources, see notes 9–11 and accompanying text.

discussed in chapter 2,[82] whatever system we choose to adopt, be it the Coasean market-based private property approach, the bureaucratic command and control approach, or a mixed system, we first have to anchor the system in a regulatory regime that defines initial entitlements and provides the necessary infrastructure for their efficient allocation. Any market-based system – say, for example, an institution that auctions allocations of fresh water, sewage water, or pollution permits – will rely to a certain extent on a regulatory regime that defines, assigns, and monitors entitlements. The transboundary resources we are discussing are resources that cannot be developed sustainably and allocated efficiently without strong support mechanisms that replace the otherwise simple enforcement mechanism of fences, boundaries, and the like and protect third party interests.

The emphasis on market-based management derives from the concern with command and control mechanisms that tend to become inefficient, slow, and captured by small interest groups. There is, of course, more than a grain of truth to this concern. But as we know from experience, there is also a market for command and control mechanisms. In other words, under certain conditions, efficiency induces potential traders to merge their activities in a joint venture. It is my contention that these conditions exist in the sphere of transnational ecosystem management, and hence, states that share such resources are more likely to establish such joint ventures than to engage in arm's-length trade.

The evidence for the existence of a market for joint ventures is ample and well analyzed. In contract theory, it is discussed in the context of the distinction between discrete and relational contracts.[83] Discrete agreements usually involve an exchange of non-specific goods, whereas relational contracts are usually long-term contracts relating to idiosyncratic goods and, thus, create relations between the parties. Cooperation between parties into the future requires much trust, as well as assurances against defection, since both parties are concerned with changing circumstances and shifts in the power relations between them throughout the life of the contract. Parties who enter a relational agreement cannot rely on the market to enforce commitments. They have no external

---

[82] See discussion at chapter 2, notes 19–22 and accompanying text.

[83] On the distinction between discrete and relational contracts, see Ian R. MacNeil, *The New Social Contract: An Inquiry into Modern Contractual Relations* (New Haven, Yale University Press, 1980). On the game-theoretical aspects of relational contracts, see Robert E. Scott, "Conflict and Cooperation in Long-Term Contracts" (1987) 75 Calif. L. Rev. 2005.

options to guarantee themselves, one in relation to the other, against deteriorating performance or changes in their relative power positions. Due to the uncertain future conditions and the inability to characterize complex adaptations, the parties, when constructing their relations, are incapable of reducing important terms of the arrangement to well-defined obligations. Most of the emphasis is therefore placed on the structures and procedures for future exchanges.[84] The core of such agreements is mutuality and flexibility. The relational agreement assigns obligations fashioned by relatively vague standards such as "best efforts" and "reasonable care" and concentrates on structures for monitoring compliance, adjusting to new circumstances, and providing internal responses to possible breaches. These undefined mutual and flexible relations are sustained within an institution that is assigned the task of providing a forum for interactions among the parties and of adapting their obligations to changes in the circumstances.

Agreements to establish transnational ecosystem institutions, especially those that include investing in infrastructures such as dams, hydroelectric facilities, and irrigation projects, epitomize the definition of relational agreements. They are, in fact, transnational relational agreements, an important – if not properly recognized and explored – category of treaties.

Therefore, the real question is what is the optimal mix of command and control and markets within transboundary ecosystem management institutions? If we want to remain faithful to the democratic vision of limited government and to the idea of the free market, we can rephrase the question to how can institutions be constructed to allow a minimal amount of central planning and intervention and as much trade as possible? To respond to this question, we must first assess the potential causes of market failure.

Perhaps the most challenging difficulty arises in the context of definition of shares. As discussed in chapter 2, the natural characteristics of water and air complicate the process of defining property rights in transboundary ecosystems. Much collective effort must be invested in this process. Chapter 1 is replete with examples of ancient communities that overcame these problems in the sphere of irrigation rights by devising complex systems of rules for defining entitlements and allowing their restricted trade, backing them up with measures to monitor compliance,

---

[84] Oliver E. Williamson, *The Economic Institutions of Capitalism: Firms, Markets, Relational Contracting* (New York, Free Press, 1985).

to resolve disputes, and to respond to acute supply crises. Many more, contemporary examples have been reported.[85]

Once shares have been defined, some restrictions on trade may be necessary. First, there may be a need to restrict out-of-basin transfers. Such a restriction was found to be necessary in ancient times. While the villagers sharing common water resources could trade water among themselves, they were not allowed to sell or lease water to farmers whose fields lay outside the perimeters of the village.[86] This rule reflects an early understanding of hydrology: out-of-basin transfers mean the loss of opportunity to reuse water within the basin. In ancient times, the out-of-basin user was usually an outsider to the group of users and was less likely to be influenced by its social norms, and hence, the group could rely less on his sharing the burden of collective action. When designing contemporary management systems, especially at the local, grassroots level, attention must be paid to these constraints.

More difficulties arise once the task of ecosystem management institutions extends beyond allocating water for irrigation. When several different users, from industry to the recreational business, from the city to the village, compete for entitlements for several uses, including drinking water, sewage, hydropower, cooling-off reactors, and recreation, market exchanges become complicated. Some demands, including the unique demands of minority groups, should be protected against small interest group capture. Basic subsistence, for example, should not be subject to market forces. Domestic supplies of water, as well as a minimum level of clean air, must be immune to the market.[87] Similarly, allocations for environmental conservation, biodiversity, and minimal instream flows to safeguard estuarine ecosystems have not fared well in some markets and, hence, should be given special care and protection.[88] Another

---

[85] See Elinor Ostrom, *Governing the Commons: The Evolution of Institutions for Collective Action* (Cambridge, Cambridge University Press, 1990).

[86] See Eyal Benvenisti, "Collective Action in the Utilization of Shared Water Resources: The Challenges of International Water Resources Law" (1996) 90 AJIL 384, 386.

[87] Bauer, "Water Markets," note 8 (describing the adverse effects of Chile's 1981 Water Code, which allowed private transactions in water backed by a weak regulatory system); Jose A. Rivera, "Irrigation Communities of the Upper Rio Grande Bioregion: Sustainable Use in the Global Context" (1996) 36 Nat. Res. J. 491 (describing the adverse effects of emerging water markets in New Mexico on the aboriginal Pueblo Indians, predicting the drying up of their villages and the collapse of their culture and discussing strategies for cultural survival).

[88] Gregory A. Thomas, "Conserving Aquatic Biodiversity: A Critical Comparison of Legal Tools for Augmenting Streamflows in California" (1996) 15 Stan. Envtl. L. J. 3 (analyzing the institutional arrangements under federal and California laws for intervention on behalf of the public interest in water trade).

complicating factor is the growing risks entailed by many uses. The uncertain effects of many of these risks add a new dimension to market failure, as market-based risk management tends to be inefficient.[89]

In the transnational sphere, market-based systems pose additional burdens. The first obstacle is the small number of participating states, the joint owners of the shared ecosystem. The smaller the number of participants, the less efficient the market. For this reason, the general call for subsidiarity[90] makes special sense in this context. Subdividing the state will increase the number of actors in the regional market. Thus, provinces, districts, local communities, as well as firms and nature conservation NGOs should be included as potential contenders for entitlements.

Another difficulty in the transnational scene derives from protectionist policies. Farmers, for whom irrigated agriculture is a way of life, might refuse to trade their water rights. Governments, who dread further urbanization and are influenced by strong domestic interest groups, might find ways to subsidize farmers' bids for an even larger share of entitlements, as well as to subsidize heavily the price of water and produce to them (up to zero cost of water). Even today, farmers remain the most successful domestic group in terms of protection against foreign trade.[91] As a result, governments often overinvest in domestic food production, and hence tend to overuse transboundary resources such as water and soil. This tendency is further reinforced by the fact that governments, particularly in developing countries with unstable supplies of water, tend to maintain a strategic interest in food security. In particular, they strive to ensure sufficient water for fresh produce and milk, which may not be easily obtained through international trade. And if imports are not a reliable source of foodstuffs, a wider margin of food security is necessary. A policy of food security in arid and semi-arid countries is wasteful, compared to a policy of enhanced trade. But uncertainties with respect to reliance on foreign sources lead many governments to pursue that inefficient policy nonetheless. Another concern for many

[89] See Breyer, *Vicious Circle*, note 41, at pp. 56–7.
[90] On subsidiarity, see chapter 6, notes 22–34 and accompanying text.
[91] On the domestic political leverage held by farmers in developed countries see Trebilcock and Howse, *Regulation,* note 32, at pp. 259–60. Similar leverage is exercised in many developing countries, as the story in Jordan (chapter 3, note 19) demonstrates. See also Henrik Horn, Petros C. Mavroidis, and Hakan Nordstroem, "Is the Use of the WTO Dispute Settlement System Biased?" CEPR Discussion Paper No. 2859 (London, 2001) (surveying contemporary trade disputes and concluding that "agriculture is still one of the most contentious areas of the WTO Agreement").

governments in developing countries that leads to protectionist policies is the impact of restricted access to water on peasants' livelihoods. For the peasants in these countries, water means employment and maintenance of their livelihood in their remote villages. Lack of sufficient water supplies would entail dislocation for them and, for the governments, demographic changes and heightened social pressures as cities become overcrowded. Thus, a transnational system of trade in water and environmental rights must provide minimal allocations as well as other assurances to those participating states concerned with food, water, and environmental security.

Some of these problems call for regulation not of the resource in question but of the trade in its derivatives. Stricter rules that minimize protectionism in the trade of agricultural products, for example, could reduce governments' concern with self-sufficiency, and hence reduce demand for precious water and land. As Tony Allen has observed, the more reliable trade in international goods, or "virtual water," is, the less there is a concern for food security.[92] Instead of irrigating fields, governments could import food and the water contained in it. Yet in agriculture, market protectionism is still the dominant policy, even in OECD countries. In fact, government support of agriculture has been on a steady increase, and efforts to change course under the auspices of the World Trade Organization's Agreement on Agriculture still face strong domestic resistance.[93] Short-term, domestically constrained policies yield global waste of strained resources.

Because the supply of natural commodities, in particular, fresh water, can fluctuate greatly in some areas and because demands also are subject to change, the transnational trading system should facilitate reallocation of entitlements. Attention must be given to the nature of the traded entitlement. There is a whole range of options: from entitlements protected by a property rule[94] – namely, entitlements that cannot be modified without the owner's consent – to entitlements that are subject to periodic reevaluation and possible modifications (with or without compensation), through to temporary entitlements that are valid only for short time periods. Entitlements could be assigned only to specific uses

---

[92] See J. A. Allan, "Overall Perspectives on Countries and Regions" in Rogers and Lydon, *Water in the Arab World* note 7, pp. 65, 68.

[93] On trade in agriculture see Trebilcock and Howse, *Regulation*, note 32, chapter 10.

[94] Following the distinction developed in Guido Calabresi and A. Douglas Melamed, "Property Rules, Liability Rules and Inalienability: One View of the Cathedral" (1972) 85 Harv. L. Rev. 1089.

of a shared resource, such as, for example, to fresh water earmarked for irrigation only (rather than for domestic uses) or reclaimed sewage water for certain agricultural or industrial purposes.

Finally, there is the issue of dispute settlement. Under the aegis of transnational institutions, disputes may arise among different users as well as between the users and the institution. Effective dispute-settlement mechanisms are a prerequisite for the definition of entitlements and, hence, for successful markets. The main reason for concern in structuring the adjudication process is that courts and arbitrators are generally ill-equipped to explore the "Pareto frontier" – namely, the most efficient decision given all the relevant data – in the context of transnational ecosystem management.[95] The process of litigation provides parties with incentive to exaggerate and misinform with regard to their domestic constraints and to refrain from exploring their differences in valuation, preference, risk aversion, and other dimensions. Even more problematic is the possibility that two parties to litigation could use the judicial process to reach a bilateral settlement that would externalize costs on third parties. Hence several procedural aids, such as the use of impartial experts or allowing standing to foreign governments and even NGOs in the litigation process,[96] should be developed to correct the deficiencies involved in third party dispute settlement mechanisms.

Conclusion: transnational ecosystem institutions as a remedy for market failures

This chapter explored the complex tasks of transboundary ecosystem management. Ecosystem management requires a constant balancing of conflicting interests and even human rights, under constraints imposed by nature and by the limited ability of humans to assess risks. To meet these challenges, it is necessary to structure the decision-making processes in transboundary ecosystem institutions in ways that will ensure informed and unbiased decisions. Such institutions could sustain

---

[95] On the relative deficiency of courts in performing these tasks in the context of ecosystem management, see chapter 7, text accompanying notes 4–14.

[96] NGOs could be brought in through their filing of amicus briefs. This option is hotly debated now in the context of the WTO's dispute settlement process. See *United States – Import Prohibition of Certain Shrimp and Shrimp Products:* Report of the WTO Appellate Body, WT/DS58/AB/R (1998) (permitting the US to submit an NGO brief as part of its brief); *European Communities – Measures Affecting Asbestos and Asbestos-Containing Products*, Communication from the WTO Appellate Body, WT/DS135/9 (8 November 2000) (inviting NGOs and others to submit applications to file briefs).

markets for some of the uses of the resource, say, for trade in pollution permits, for water for irrigation, or for reclaimed sewage water intended for agriculture, provided the institution can ensure the attainment of the tasks of resource-, claims-, and risk-management and can comply with the external normative constraints discussed in this chapter. In light of those tasks and constraints, the next chapter examines the principles and procedures according to which transnational ecosystem institutions should be designed.

6   The structure and procedure of institutions for transboundary ecosystem management

Introduction

This chapter discusses the different characteristics of the transnational institutions for ecosystem management required for responding to the challenges identified in the previous chapter. Shared management of transnational resources requires sensitivity to the interaction between the shared institution and the participating national governments. A carefully planned system of checks and balances must be created to prevent ineffective joint management, on the one hand, and inattention to national concerns, on the other. States will agree to confer sovereign authority to the joint institution only if they are allowed to retain important tools – such as veto power over decisions, control of the institution's budget, sufficient representation in the institution's bureaucracy, and judicial review over the institution's decision – to ensure them reasonable control over the institution's decision-making processes.

In light of these requirements, the first part of this chapter discusses the structure of transnational ecosystem institutions, addressing such issues as subsidiarity, the relationship between the transnational and the national processes, and the relationships between different institutions with overlapping competences. Emphasis is placed on the possibilities for reducing the likelihood of skewed or uninformed decisions.

The second part examines the decision-making processes within transnational institutions, focusing on the effort to provide flexibility and mutuality in a transparent process that ensures voice to the interested public.

The structure of transnational ecosystem institutions

This part addresses the relationship between the national and the transnational levels. The argument is that the transnational decision-making processes must be given supremacy over national policies to ensure effective management of the shared ecosystem. This necessarily entails an appreciable loss of sovereignty for the national governments and loss of voice for citizens. One response to these concerns is the delegation of institutional power to sub-basin units, the smaller parts of the ecosystem whole. Such subsidiarity provides voice to citizens residing in the different sub-basins and a role for the local governments.

Supremacy of institutional policies

National policies and procedures affect the possibility of regional cooperation with regard to shared air, fresh water, and other natural resources. The national legal and institutional arrangements for the domestic allocation and monitoring of the uses of such resources shape each state's ability to commit itself to international obligations and to comply with them. These domestic policies and institutions are relevant in a number of ways. First, the method for allocating shares among individual users, ranging from a rigid system of inalienable property rights to a flexible system of revocable permits, impacts the government's ability to undertake to implement a reduction of its share of a transboundary natural resource. The existence of property rights to the resource may tie the hands of state negotiators, willingly or unwillingly, or increase the enforcement costs due to litigation of expropriation cases. In contrast, a revocable, permit-based system provides more leeway for decision-makers.[1] Second, different internal allocation methods shape differently the incentives for users to intervene in the political process. The more rigid the allocation system, the greater users' reliance on their "property rights," and hence, the greater users' incentives to invest in protecting those rights through obstruction of an international agreement. Finally, poor administration and ineffective monitoring of uses and users by the government may further burden the difficult task of implementing the international undertakings or be used as an excuse

---

[1] For a similar suggestion in the context of US state law, see Joseph W. Dellapenna (ed.), *The Regulated Riparian Model Water Code: Final Report of the Water Laws Committee of the Water Resources Planning and Management Division of the American Society of Civil Engineers* (New York, American Society of Civil Engineers, 1997), p. 200.

for failure to comply with them. Not surprisingly, powerful domestic groups are usually behind the existence of rigid allocations and poor governmental controls.

The operation of transnational institutions, which must be based on the principles of flexibility and mutuality, must not be constrained by the methods used by the participating states in the domestic allocation of their shares of the transboundary natural resource. This implies two further principles: first, the policies of transnational institutions must enjoy supremacy over domestic policies; and second, transnational institutions must have the competence to dictate changes in domestic law relevant to the management of the transboundary resource. The legal precedence given to institutional policies implies, first, that these policies will take effect within the national legal systems without a need for securing prior ratification from the national legislatures or governments as though each policy is a new treaty in itself. Second, supremacy of institutional policies also requires states to modify their domestic legislation regarding resource use in order to enable transnational policies to take effect. One important implication of the principle of supremacy is that instead of rigid systems that provide owners with inalienable property rights in specific shares of a transboundary resource, each participating state must establish a flexible system of revocable permits for individual users of the resource.[2] Although governments are usually empowered to take private property and, hence, can also take property rights in shares of a resource from their owners, the process of taking, especially when protected by constitutional guarantees and judicial scrutiny, is more complex and expensive than the termination or non-renewal of temporary permits.

Such a flexible, permit-based system is vital to transnational cooperation for three reasons. First, it is a prerequisite for the regional management of transboundary resources, which must remain flexible in order to endure.[3] Second, a permit-based allocation system requires an institutional framework that assigns, amends, and revokes permits. Such an institution could lower the likelihood of skewed domestic allocations to powerful groups of users by providing procedural guarantees for accountability in decision making. Finally, a permit-based system and its institutions encourage equal respect for the demands of all users,

[2] The elimination of allocation of shares protected by property rights does not preclude trade in such allocations. There can be, for example, markets for revocable permits issued periodically by the managing institutions.

[3] See the discussion in the text accompanying notes 35–8.

because they have to base allocative decisions on notions of basic human rights and equal access to national resources.[4]

The supremacy of transnational management institutions implies that the locus for interest group activity shifts from the domestic political scene to the transnational institution. The risk of interest group capture at the institution level is substantial. Hence, the decision-making process within the transnational institutions will have to establish guarantees to contain this risk.[5]

Supremacy of institutional policies entails reduced opportunities for the state parties to exit from their treaty obligations. A reduction in exit opportunities increases the commitment of all parties to cooperation. States may seek to exit as a result of pressure from interest groups or due to hostile public opinion manipulated by interest groups or terrified by perceived new health or other risks. A number of doctrines are included in the arsenal available to governments seeking to exit, among them: the right to "terminate" a treaty in retaliation for another party's prior material breach;[6] the claim of a "state of necessity;"[7] and the doctrine of *rebus sic stantibus*, which allows a party to unilaterally withdraw from its treaty obligations if it can show that "a fundamental change of circumstances has occurred with regard to those existing at the time of the conclusion of the treaty, and which was not foreseen by the parties."[8] Lowering the threshold for such claims would increase the likelihood of domestic pressure being brought to bear on governments to invoke them and unilaterally exit from international obligations. Because long-term cooperation is the key to sustainable and optimal resource management and because the management of natural resources is particularly susceptible to domestic pressure to renege, a high threshold for exerting a unilateral right of exit is in the long-term interest of the participating states and, therefore, desirable. Thus, all things being equal, the stricter the rules precluding unilateral exit from treaty obligations, the stronger

---

[4] On the human right to the environment, see chapter 5, text accompanying notes 57–79.

[5] See the discussion in the text accompanying notes 23, 32, 50.

[6] Article 60 of the 1969 Vienna Convention on the Law of Treaties.

[7] For a discussion of this claim, which was raised by the Hungarian government, see International Court of Justice, *Case concerning the Gabcikovo–Nagymaros Project (Hungary/Slovakia)* (1997), Judgment, ICJ Reports 1997, p. 7, reprinted in http://www.icj-cij.org/idocket/ihs/ihsjudgement/ihsjudframe1.htm; (1998) 37 ILM p 167 at para. 104.

[8] Article 62(1) of the 1969 Vienna Convention on the Law of Treaties. See also the *Gabcikovo–Nagymaros Case*, note 7, paras. 49–59.

a state party's commitment to long-term cooperation, and the lower the uncertainty as to possible future breaches.

From this perspective, the International Court of Justice came to the right conclusion in its 1997 decision in the *Gabcikovo–Nagymaros Project* case.[9] The decision clearly follows the logic of raising the hurdle for state parties attempting unilateral exit. In general, the unilateral exit option from treaty obligations concerning transboundary resources is quite limited. In most cases, the change in supplies or demands from the resource would be incremental, fairly foreseeable by the parties to the agreement over the initial allocations and, hence, would not amount to "fundamental change of circumstances." But this dispute did not involve regular circumstances. The project under scrutiny was of huge proportions and irreversible consequences, conceived by communist regimes from a long-gone era, with little attention to the environment or to sustainable development. The agreement signed in 1977 between the Hungarian People's Republic and the Czechoslovak People's Republic provided for the construction and operation of a system of locks on the Danube River, between Gabcikovo (in Czechoslovakian territory) and Nagymaros (in Hungarian territory), which would provide for diversion canals and two hydroelectric power plants. The project was to be financed, constructed, and operated jointly, on an equal basis between the two countries. The dispute between Hungary and Czechoslovakia (and, later, with Slovakia) with regard to the project led both sides to resort to unilateral moves. Hungary first suspended its implementation of the treaty between the two states and subsequently declared its termination, while Czechoslovakia constructed a canal to divert the Danube and then put the canal into operation by damming the river. Nevertheless, the court rejected these moves as having no effect on the 1977 treaty and stipulated that unless changed through bilateral consent, both parties are bound by it. This ruling came despite the fact that most of the judges approved the unilateral moves of one side or the other and only six judges rejected all of the unilateral acts.[10] The treaty had survived the momentous political changes in both countries, and the court required the two parties to negotiate in good faith the promotion of the objectives of the treaty and to deepen their cooperation to include the establishment

[9] See note 7.
[10] Five judges approved Czechoslovakia's implementation of the provisional solution, whereas four other judges approved Hungary's termination of the treaty (note 7, para. 155).

of a "joint regime."[11] In reaching this conclusion, the court deliberately emphasized international undertakings at the expense of domestic pressures. It rejected Hungary's claim that a "state of ecological necessity," even if it existed, outweighed the wrongfulness of its unilateral suspension of the project, because Hungary could resort to negotiations to reduce the environmental risks.[12] It similarly rejected Hungary's claims to impossibility of performance, fundamental change of circumstances, and lawful response to Czechoslovakia's earlier material breach (namely, Slovakia's construction of the provisional diversion project).[13] The ICJ was critical also of Slovakia's moves. It found the Slovak diversion of the Danube waters a breach of its obligation to Hungary to respect the latter's right to an equitable and reasonable share of the river.[14] Despite its findings that both sides had failed to comply with their obligations under the treaty, the court concluded that "this reciprocal wrongful conduct did not bring the Treaty to an end nor justify its termination."[15] Finding the agreement flexible and, therefore, renegotiable, the ICJ imposed the 1977 treaty on both sides, ordering them to negotiate its implementation while taking into account current standards on environmental protection and sustainable development[16] and to regard Slovakia's diversion dam and canal as a "jointly operated unit" under the treaty regime.[17]

Without entering into the doctrinal aspects of this judgment,[18] its implications for the interface of domestic and international politics are illuminating. The judgment clearly seeks to shield international politics from the influence of domestic politics (and perhaps also provide what they perceive as an appropriate ad hoc solution, noting the less than catastrophic outcomes of the "provisional solution" as implemented by Slovakia). Despite the momentous internal political, economic, and social changes in both countries and the relentless public pressure and even parliamentary resolutions urging the Hungarian government to stop the project, domestic options remained constrained by an agreement conceived in a long-gone era. Even when one side breaches

---

[11] *Gabcikovo–Nagymaros Case*, paras. 144–7.

[12] *Ibid.*, para. 57 (it found that Hungary did not prove an imminent peril, but, rather, uncertain assumptions concerning long-term damage).

[13] *Ibid.*, paras. 101–12.   [14] *Ibid.*, para. 78.   [15] *Ibid.*, para 114.

[16] *Ibid.*, paras. 138–40.   [17] *Ibid.*, para. 146.

[18] The decision contains a number of important developments to international environmental law and freshwater law. On inter-generational equity, see chapter 5, text accompanying notes 45–56. For a thorough analysis of the decision, see chapter 7.

its obligation to renegotiate in good faith, the government of the other side cannot succumb to internal public pressure and take unilateral moves: it must exhaust all possible means, even third-party intervention, to persuade its partner to return to the negotiating table. The ICJ closes another potentially crucial loophole in restricting the ability of a successor state to sidestep treaties concerning transboundary resources ratified by its predecessor. Treaties relating to shared resources, such as the 1977 Czechoslovak–Hungarian treaty, are considered "localized Treaties" and, therefore, survive state succession.[19]

The court clearly stated its preference for strong regional joint management institutions, institutions that "reflect in an optimal way the concept of common utilization of shared water resources."[20] It undoubtedly envisioned that in future disputes of this kind, governments, armed with this instruction from the ICJ, would be able to dodge domestic opposition to an international agreement simply because any alternative government would be likewise bound by the obligations.

This outcome of limited exit options, as demonstrated in the *Gabcikovo–Nagymaros* case, has its price. Assuming that governments generally put their short-term interests first, existing agreements will tend to reflect the preferences of the stronger domestic actors at the time of signature. Often, these actors will be polluters and heavy users who would opt for lax control of their uses and for imposing externalities on third parties. Hence, domestic interest groups that subsequently gain power will be constrained by what might be deemed the undesirable "fatal embrace" of their predecessors. Indeed, the 1977 treaty serves as a prime example of an agreement whose implementation could bear dire future consequences for the shared environment and the communities dependent upon it.

The best way to overcome this difficulty is by treating such agreements as flexible frameworks and subject to evolving international law, which, in turn, is influenced by new understandings of sound management of natural resources. The ICJ pursued this route by reading into the 1977 treaty a flexibility that opened the way to renegotiating its basic provisions in light of new developments in international law, new understandings of environmental impacts, and new circumstances. Indeed, the court upheld the 1977 treaty only after reading into it flexibility and

---

[19] See Article 12 of the 1978 Vienna Convention on Succession of States in Respect of Treaties, which was recognized by the ICJ as reflecting customary law and applied to the 1977 treaty in the *Gabcikovo–Nagymaros* case (note 7, paras. 122–3).

[20] See note 7, para. 147.

mutuality and emphasizing the obligation of both parties to achieve the object and purpose of the developing treaty relationship (an object and purpose that the court, in fact, postulated in light of developing international law).[21]

The conclusion of this part must not overlook the downside of transferring competences from national governments into the hands of transnational institutions. Delegating sovereign authority to ecosystem management institutions reduces dramatically the powers of all participant states. This reduction is dramatic not only in the narrow context of the allocation of the specific resource, but also in many relevant ancillary contexts. Because, for example, diverse users and uses can affect the supply of clean and ample water, control of activities that can potentially impact water availability implies intervention in a considerable chunk of national regulation, involving different aspects of life and branches of government, to ensure compliance with the policies of the transboundary resource institution. In return for such a broad delegation of authority, governments would insist on maintaining control over the decision-making process within the institution and over its implementation processes. There are various ways to ensure such control, from requiring decision by consensus, to control of the institution's budget, through to judicial appeal or review procedures of institution decisions. These tools will be discussed in the next part below.

Integration cum delegation of authority: the case for subsidiarity

A potentially effective way to overcome the tension between the transnational institution and the participating national governments is to create links between the regional institution and sub-state entities such as provinces and towns. We saw in chapter 3 that the constitutional design of a government as a clearinghouse for the diverse and conflicting domestic interests is often responsible for the failure to reach agreements on transboundary resource management. The monopolistic position of the government requires that any domestic actor, such as a local municipality or provincial government, seeking to establish cooperation across international boundaries with neighboring sub-actors must invest resources to persuade the government to represent the actor vis-à-vis the other government. Either of the relevant national governments may,

---

[21] See note 7, paras. 132–47. Judge Bedjaoui criticized this evolutionary interpretation; see his separate opinion at note 7; http://www.icj-cij.org/idocket/ihs/ihsjudgement/ihsjudframe1.htm.

however, have different interests from the domestic actor seeking their representation, due to the influence of other domestic sectors or the demands of other issues. To overcome this structural failure, it is necessary to develop the opportunity for direct, low-level interaction among sub-state actors.

Lower-level decision making and interaction could serve additional goals. From the perspective of efficiency, lower-level interaction can increase regulators' understanding of the particular natural attributes of a local resource and the potential impacts on it by the suggested policies. As Daniel Esty argues, "Bureaucrats in Washington . . . cannot know the future land use of a contaminated waste site as well as those in the community where the site is located. In deciding 'how clean is clean enough,' local judgment is essential."[22] Capture by interest groups may be less effective in local settings, whereas public participation could be less costly and more capable of influencing outcomes in conformity with the public good.[23] Public participation may also have a positive influence on locals' commitment to compliance.[24] The existence of numerous sub-basins may provide the foundation for an efficient market between sub-basin institutions.[25] Additional support for delegation of authority to sub-state levels derives from the perspective of democracy: delegating authority to local institutions increases the opportunities for citizens to take an active part in influencing their lives.

These considerations of efficiency and democracy can be further bolstered by considerations of human rights and group rights. As mentioned in the previous chapter, there is a growing recognition in international law towards respecting and promoting the claim of minority groups, especially indigenous peoples, to the right to autonomous management of natural resources in their vicinities as part of the claim for self-determination and protection of their cultures.[26] Delegating

[22] Daniel C. Esty, "Revitalizing Environmental Federalism" (1996) 95 Mich. L. Rev. 570, 625; see also Daniel B. Rodriguez, "The Role of Legal Innovation in Ecosystem Management: Perspectives from American Local Government Law" (1997) 24 Ecology L.Q. 745 (local government law and policy represent a largely untapped opportunity to pursue sound regulatory strategies alongside state and federal efforts).

[23] Michael C. Dorf and Charles F. Sabel, "A Constitution of Democratic Experimentalism" (1998) 98 Colum. L. Rev. 267, 314–23; Timothy P. Duane, "Community Participation in Ecosystem Management" (1997) 24 Ecology L.Q. 771; Neil A. F. Popovic, "The Right to Participate in Decisions that Affect the Environment" (1993) 10 Pace Envtl. L. Rev. 683.

[24] Esty, "Revitalizing," note 22; Daphna Lewinsohn-Zamir "Consumer Preferences, Citizen Preferences, and the Provision of Public Goods" (1998) 108 Yale L. J. 377, 398.

[25] On this subject, see chapter 5, notes 80–96 and accompanying text.

[26] On this subject, see chapter 5, notes 9–11, 78–9 and accompanying text; chapter 7, notes 133, 144–5 and accompanying text.

authority over transboundary ecosystem management may, therefore, be beneficial not only economically, but also socially.

The idea of delegating the task of natural resource management to sub-state agencies has been tested with much success in various countries.[27] In the transnational arena as well there are ample examples of sub-state transboundary cooperation in matters related to environmental protection and water utilization, particularly in Western Europe, but also between US states and Canadian provinces, and between American and Mexican cities.[28] These examples indicate that in order to reduce the complexity entailed in the interaction between such heterogeneous actors as states, it may be beneficial to resort, at times, to sub-state agreements negotiated by sub-state actors, such as governors of states or mayors of neighboring cities.

Sub-state cooperation may be particularly vital when national interests such as security or trade overshadow resource politics. A recent case in point is the agreement signed in 1996 between two municipalities, the Regional Council of Emek–Hefer, located on the coastal plain of Israel, and the municipality of Tul-Karem, in the Palestinian-controlled area of the West Bank. These two municipalities share a severely polluted small basin, in which run-off from Palestinian towns and villages as well as from Jewish settlements flows via a small stream across the Green Line into the Emek–Hefer region. Since negotiations between the Israeli government and the Palestinian Authority were blocked, the only avenue open to local administrators for pursuing rehabilitation was direct low-level negotiation, thereby bypassing the deadlock at the national level. After receiving an implicit green light from both governments – the

---

[27] For a survey of sub-state agencies in the US, France, and New Zealand, see Lloyd Burton and Chris Cocklin, "Water Resource Management and Environmental Policy Reform in New Zealand: Regionalism, Allocation, and Indigenous Relations" (1996) 7 Colo. J. Int'l Env. L. & Pol'y 75.

[28] Joachim Blatter and Helen Ingram, "States, Markets and Beyond: Governance of Transboundary Water Resources" (2000) 40 Nat. Res. J. 439, 448–53, 456–8; Maria Teresa Ponte Iglesias, "Les accords conclus par les autorités locales de différents Etats sur l'utilisation des eaux frontalières dans le cadre de la coopération transfrontalière" (2/1995) Schweizerische Zeitschrift fuer Internationales und Europaeisches Recht, 103, 129–30; Ulrich Beyerlin, "Transfrontier Cooperation between Local or Regional Authorities," Encyclopedia of Public Int'l L. (Installment 6) 350; Pierre-Marie Dupuy, "La coopération régionale transfrontalière et le droit international" (1997) 23 Ann. Franc. Dr. Int'l 837. See also the New York–Quebec Agreement on Acid Precipitation (1982), reprinted in (1982) 21 ILM 721, and the 1994 agreement between national and regional governments concerning the protection of the rivers Meuse and Scheldt (reprinted in (1995) 34 ILM 851).

Israeli Ministry of the Environment and Chairman Arafat – the heads of the two municipalities met and signed an agreement outlining their mutual commitment to cooperation.[29]

A similar case was reported in the former Yugoslavia, during the bloody 1991–2 conflict.[30] Despite the atrocities of the inter-ethnic conflict and the many incidents of intentional destruction of dams, a low-level agreement was reached in 1992 between the Serbs controlling the upstream Trebisnica River in Bosnia-Hercegovina and the Croat managers of the Dubrovnik hydropower plant. The agreement permitted the continuous flow of the river to the Dubrovnik plant in exchange for the Croats' guarantee to allow continuation of the supply of the river's waters to the Bay of Kotor area in Montenegro.

The last example is that of the city of Nicosia. This city is torn apart by Greek and Turkish Cypriots. Nevertheless, it has only one, integrated sewage system, constructed before the division of the city. Despite the deep hostility between the two groups, the sewage system remains operative, maintained through low-key joint operation of technical teams from both sides.[31]

[29] The originals are in both Arabic and Hebrew. They have not been published. My translation is as follows:

> *Letter of Intent*
> The District of Tul-Karem, the Tul-Karem Municipality, and Emek–Hefer Regional Council recognize the acute necessity to promote and protect the environment, for the protection of the water we drink and the soil we cultivate, for the benefit of the inhabitants of Tul-Karem and environs, the Hefer Valley and environs.
>
> It was therefore decided to establish a steering and planning committee that will be entrusted with supplying mutual expert solutions to resolve the problems in the short and immediate range and in the long range.
>
> Those who stand at the helm will jointly work to obtain funding and consent from international bodies, in an effort to realize the plans and to implement them.

> The written text, in both languages, was prepared in advance by Mr. Itzkovich. He was accompanied by Mr. Abu-To'ama, the mayor of an Arab municipality in Israel, who made the initial contacts. Mr. To'ama also signed the letter. The envisioned plans are rather ambitious and complex and include sewage-treatment facilities to be constructed with international financing on West Bank territory, to supply the treated water for Palestinian agricultural use.

[30] Mladen Klemencic, "The Effects of War on Water and Energy Resources in Croatia and Bosnia," in Gerald H. Blake *et al.* (eds.), *The Peaceful Management of Transboundary Resources* (London, Graham & Trotman, 1995).

[31] For options for low-key yet crucial joint ventures for Jerusalem and a description of the Nicosian model, see Eran Feitelson and Qasem Hassan Abdul-Jaber, *Prospects for Israeli–Palestinian Cooperation in Wastewater Treatment and Re-use in the Jerusalem Region* (Jerusalem, Institute for Israel Studies, Palestinian Hydrology Group, 1997).

The promise of sub-state cooperation is of relevance at the stage of designing joint institutions for transboundary ecosystem management. Instead of relying on member states as the basic building blocks of such institutions, these institutions could be based on a system of smaller sub-units that coordinate the use of the resource in the different sub-components of the ecosystem. Thus, for example, instead of a river commission headed by representatives of all participating national governments, the system could alternatively be based on a cluster of sub-basin institutions, each comprising representatives of the local communities in each sub-basin. The existence of a number of smaller institutions, each responsible for a single sub-basin, could facilitate efficient intra- and inter-basin trade in shares of the resource.[32] The higher institution could serve as a forum for negotiations and even a clearinghouse for transactions among sub-basin representatives. In politically sensitive areas, such small-scale de-politicized ventures could prove the only plausible means of cooperation.

This idea of a special type of subsidiarity is not a panacea, and there may be very good reasons why it should be resisted partly or entirely. Economies of scale suggest that questions of risk assessment and even risk management should be explored on the international level. Interest group capture may at times be stronger at the local level than in the international sphere.[33] There could be regions where for social and economic reasons, cross-border cooperation is particularly difficult, and thus, cooperation at the national level would be more effective. In multi-ethnic countries, central governments may be worried that cross-border cooperation will spur secessionist efforts by ethnic minorities situated in border areas. But these concerns do not rule out the promise of localized cross-border cooperation agreements where they can produce positive results. These agreements require a refined approach to the design of the level of natural resource regulation.

The design of frameworks for the operation of subsidiary cross-border institutions must rely on a clear normative basis. When I asked the Mayor of the Emek–Hefer Regional Council, Mr. Itzkovich, what he thought was the legal status of the inter-municipal agreement he had signed with the Governor of the Tul-Karem District, he did not have

---

[32] See Eyal Brill, "Applicability and Efficiency of Market Mechanisms for Allocation of Water with Bargaining" (Ph.D. dissertation, submitted to the Senate of The Hebrew University of Jerusalem, 1997, in Hebrew), p. 102.

[33] See Susan Rose-Ackerman *Corruption and Government: Causes, Consequences, and Reform* (Cambridge, Cambridge University Press, 1999), pp. 149–50.

a clear answer. But he did say that they were both contemplating establishing an "international corporation" to take over the design and implementation of the rehabilitation plans, and for this purpose, they were seeking legal advice on the appropriate steps that must be taken.[34] Needless to say, the lack of a legal infrastructure for such undertakings may very well undermine this joint venture. This story underlines the necessity to provide a clear legal basis for the operation of such undertakings. Because such sub-state agreements cannot be based on the domestic law of one or the other states, it is necessary to develop international norms to sustain such localized cross-boundary agreements and encourage their development.[35]

Decision-making procedures

Thus far, we have discussed the structural aspect of transnational ecosystem institutions and their relationship with the national legal systems. This part will discuss the basic principles that the decision-making process within the institution must adhere to: flexibility and mutuality; the provision and analysis of information; public participation; control of the institution's agenda; and review of the institution's decisions. These principles are designed to secure two interrelated goals: to increase state parties' reliance on the institution and each other and to reduce the propensity of governments to adopt short-term goals as a result of special interest influence. In other words, the objective is to reduce the costs of collective action.

These principles are only the very basic ones for the structuring of transboundary ecosystem institutions. When designing each specific institution, states should consider a number of more detailed rules concerning its operation. These should include rules of procedure regarding the budget, size, and makeup of the institution's bodies; procedures for approving unilateral or joint policies of the participating states and non-state entities; voting rules (unanimity or majority); and rules on the nature of the institutional decisions (ranging from findings of fact, to

---

[34] Personal interview with Mr. Itzkovich, Emek–Hefer Regional Council, 21 Jan. 1997.

[35] See Agenda 21, chapter 18, reprinted in Nicholas A. Robinson (ed.), *Agenda 21 & the UNCED Proceedings* (6 vols., New York, Oceana Publications, 1992–3), vol. IV, principle 18.12 (o)(i) (recommending, "as appropriate," to develop and strengthen mechanisms at all levels concerned, including "at the lowest appropriate level" and "the decentralization of government services to local authorities, private enterprises and communities"). See also Dupuy, "La coopération régionale," note 28, at p. 860.

recommendations, through to decisions that bind the member states). These rules should include normative guidelines as to the weight to be given to conflicting considerations, such as, for example, the balance between existing and potential uses or between development and conservation. Finally, the question of whether to provide opportunities for trading in shares also should be dealt with as a specific feature of the particular institution. Many of these details depend on the specific characteristics of the ecosystem in question and on the parties sharing it. Moreover, for the reasons explored below, institutions that are based on the basic principles discussed in this chapter can function successfully even without power to bind governments or enforce decisions.

Flexibility and mutuality

When designing institutional arrangements, emphasis should not be placed on minutely defined and rigid obligations, such as, for example, with regard to allocation of quantities of water or of permitted emissions. Due to the uncertainty with regard to future conditions and the inability to foresee complex adaptations, the parties, when constructing the joint institution, are incapable of reducing important terms of the arrangement to well-defined obligations.[36] The greatest attention therefore, should be directed at structures and procedures for future exchanges. Moreover, flexibility in the institutional design is also important.[37] This observation, derived from the theory of collective action, conforms to the theory of relational contracts that distinguishes between discrete and relational contracts, discussed in chapter 5.[38] As relational contracts theory suggests, the regional cooperation agreement should be designed so as to maintain mutuality and flexibility in the relations between the parties. More specific obligations should be decided upon by the institutions to which the agreements assign governance.

Flexibility in the context of transboundary resource management implies that allocation decisions will be subject to periodic amendments in light of new conditions or knowledge that arises. This is particularly

---

[36] Andrew Hurrell and Benedict Kingsbury, "Introduction" in Hurrell and Kingsbury (eds.) *The International Politics of the Environment: Actors, Interests, and Institutions* (Oxford, Clarendon Press, 1992) (flexibility because knowledge develops over time).

[37] See Barbara Koremanos, Charles Lipson, and Duncan Snidal, *Rational International Institutions* (Rational International Institutions Project, at http://www.harisschool.uchi, 1998) (suggesting that two kinds of flexibility are necessary: flexibility of the norms and of the institutional procedures to enable the institution to modify its work).

[38] See chapter 5, notes 81–2 and accompanying text.

important in the sphere of freshwater management. Adjustment of shares in fresh water is often necessary because relative demands for water change constantly, reflecting economic and social developments in the member states, while the supply side also fluctuates with unpredictable droughts or floods. The flexible standard of "equitable and reasonable use," the core standard for allocating water resources under international law, should, thus, be understood as permitting reallocations during the lifetime of the agreement, without demanding that the party seeking modification of the allocations resort to the strict doctrine of *rebus sic stantibus*.[39] The application of the standard of "equitable and reasonable use" means that all allocations are subject to future adjustment: whenever an allocation becomes inequitable or unreasonable, the standard mandates reallocations adjusted to the new circumstances. This standard does not assign property rights in water shares, but, rather, rights subject to reevaluation that is based on objective criteria. When renegotiating allocations, existing beneficial uses are granted only a qualified priority. This standard helps to ensure flexibility and mutuality among the riparians in their future interactions and, hence, creates incentives for the parties to undertake long-term commitments and to cooperate.

Data collection, analysis, and dissemination

The key to successful fulfillment of the tasks of transnational ecosystem institutions is their ability to collect, analyze, and disseminate information concerning the performance of state and non-state parties, the conditions of the resource, the risks involved with present uses, and the available alternatives. Robert Hayton and Albert Utton, two leading experts on international water resources law in the twentieth century, have observed with regard to the management of international water resources that "there can hardly be anything more important in effecting international water resources management than the factual basis required for rational decision making."[40] Accurate and comprehensive assessment of information is a crucial component of the institution's base of legitimacy. As Hayton and Utton suggest, "Respect for the [water institution] will be rooted, in the first instance, in its thorough understanding of the circumstances of each problem. Only

[39] On this doctrine, see note 8 and accompanying text.
[40] Robert D. Hayton and Albert E. Utton, "Transboundary Groundwater: The Bellagio Draft Treaty" (1989) 29 Nat. Res. J. 663, 688–9.

in this way can it achieve impartiality in assessing the information and data it compiles."[41] Provision and dissemination of information, as well as reliance on scientific findings, can ensure the institution's accountability. Furthermore, accurate and comprehensive information enables governments to assess one another's compliance with the obligations under the agreement. In addition, the dissemination of information enables the general public to monitor the performance of its government and reduce the latter's opportunity to cater to special interests. Such a transparent decision-making process can foster domestic public deliberation within all the participating countries regarding the range of options available to the governments, thereby increasing the governments' ability to assess public support and, at the same time, constraining attempts to deviate from long-term national interests.

Shared institutions should, therefore, accumulate and provide "the widest exchange of information"[42] with regard to each member state's current and expected supplies of and demands from a shared ecosystem, as one means to ensure effective communication among state actors as well as effective monitoring by NGOs and the public at large. They should assess potential risks of existing and alternative practices and provide the basis for enlightened debate.

A similar case could be made in favor of recognizing a duty to employ and consult experts. Scientists from various disciplines, identified by some as "epistemic communities,"[43] could suggest alternatives for optimal solutions. Their contributions could diffuse politically skewed positions of domestic interest groups and, hence, of governments. Third parties such as NGOs could also provide expert advice. As the ICJ noted in its decision in the *Gabcikovo–Nagymaros* dispute, "The readiness of the Parties to accept [third party] assistance would be evidence of the good faith with which they conduct bilateral negotiations."[44]

Transparency also requires reasoned decisions. The process of reasoning and persuasion that precedes an actual vote on policy decisions and subsequently appears in the published decision is effective

---

[41] Hayton and Utton "Transboundary Groundwater," p. 689.

[42] Article 6 of the 1992 Helsinki Convention on the Protection and Use of Transboundary Watercourses and International Lakes, reprinted in (1992) 31 ILM, 1312.

[43] See Peter M. Haas, "Introduction: Epistemic Communities and International Policy Coordination" (1992) 46 Int'l Org. 1.

[44] Paragraph 143 of the decision, note 7. The court was referring to the assistance and expertise offered by the Commission of the European Communities to settle the dispute.

in eliminating inefficient outcomes and providing for more equitable distribution of resources. Such a deliberative process legitimizes the decisions taken and thus ensures greater compliance.[45] At the very minimum, the requirement that transnational institutions supply reasons for their decisions increases the accountability of the decision-makers similarly to how the reasoning requirement for court opinions serves to constrain judicial power.

Finally, when disputes arise between state parties or between a state party and the institution, special inquiries by the institution or by a special fact-finding commission within the institutional structure can prove an effective deterrent against defection. Enforcement through judicially supported sanctions is an extremely cumbersome process for state parties to shared institutions and is thus counterproductive.[46] Because enforcement is a costly collective-action problem, it tends to be underproduced as parties tend to take a free ride on the sanctioning state's back, usually the strongest of them all. Instead, finding factual evidence of a state's noncompliance, with the reputational costs it entails, may prove sufficient deterrence.[47] For this reason, enforcement and sanctions are not included on the list of the crucial components of a transnational ecosystem institution.[48] In contrast, the provision of information with regard to defectors is a vital component, along with the other information-related functions of the institution.

Public participation

Both when negotiating the establishment of joint institutions and when operating within such institutions, the acoustic separation between

---

[45] See James D. Fearon, "Deliberation as Discussion" in Jon Elster (ed.) *Deliberative Democracy* (Cambridge, Cambridge University Press, 1998), p. 44, 56. On the benefits of deliberation, see generally the contributions in Elster, *Deliberative Democracy*; Joseph M. Bessette, *The Mild Voice of Reason: Deliberative Democracy and American National Government* (Chicago, University of Chicago Press, 1994); Cass R. Sunstein, *The Partial Constitution* (Cambridge, MA, Harvard University Press, 1993).

[46] See, e.g., Abram Chayes and Antonia Handler Chayes, *The New Sovereignty: Compliance with International Regulatory Agreements* (Cambridge, MA, Harvard University Press, 1995).

[47] Chayes and Chayes, *The New Sovereignty*, p. 111; Oran R. Young, "The Effectiveness of International Institutions: Hard Cases and Critical Variables" in James N. Rosenau and Ernst-Otto Czempiel (eds.), *Governance without Government: Order and Change in World Politics*, Cambridge Studies in International Relations, 20 (Cambridge, Cambridge University Press, 1992), pp. 160, 176–8.

[48] In general, adjudication is a rather problematic technique for settling disputes related to ecosystem management. For discussion of this point see chapter 7, text accompanying notes 4–14.

negotiators and the public at large is susceptible to exploitation by special interest groups. Governments, taking advantage of the relative secrecy of international negotiations, often find it quite easy to pursue partisan, short-term policies at the expense of larger constituencies. As a result, inter-governmental negotiations often yield agreements that are skewed, sub-optimal, and unsustainable.[49] As suggested earlier, ensuring easy access to information and its unrestricted dissemination is pivotal in preventing such an outcome. But the provision of information is not enough to bridge the acoustic separation. A more meaningful way would be to allow the general public active participation in the decision-making process. A need has, therefore, been expressed to allow representation of the "other voices" in the negotiation process, primarily representatives of small communities directly affected by certain uses of the shared ecosystem or NGOs that represent the larger domestic groups that are incapable of making their voices heard. Such direct involvement can provide an opportunity for representatives of less-organized interest groups to have their concerns presented and examined not only by their governments, but also by the domestic groups of the other negotiating states. This opportunity may lower the cost of communication between environmentalists across national borders and increase their effectiveness.

Once the agreement has been ratified by the participating states and the shared ecosystem management institution established, public involvement in the ongoing decision-making processes of the institution, through consultations, hearings, or even sharing in the actual decision making, is instrumental for similar reasons. Public participation in institutional decision making has been widely recognized as crucial for responsible decision making. It has been observed that NGO participation improves the work of environmental decision-making bodies.[50] In addition to monitoring against interest group capture, NGOs provide useful information to decision-makers and otherwise contribute to improving

---

[49] On this tendency, see chapter 3.

[50] Kevin Stairs and Peter Taylor, "Non-Governmental Organizations and the Legal Protection of the Oceans: A Case Study" in Hurrell and Kingsbury, note 36, at p. 110 (describing the contribution of environmental NGOs in the development and implementation of international agreements on environment protection); Kal Raustiala, Note, "The 'Participatory Revolution' in International Environmental Law" (1997) 21 Harv. Envtl. L. Rev. 537 (describing NGOs as "major actors in the formulation, implementation, and enforcement of international environmental law" and arguing that states benefit from NGOs' informational and legitimization services).

the quality of decisions.[51] In fact, the managing institution itself is likely to promote the involvement of non-governmental voices so as to obtain relevant information and overcome governmental resistance to its policies. For example, the US–Canada International Joint Commission, the body charged with overseeing the two countries' shared water resources since 1909, has recognized that "the challenge becomes increasingly one of engaging public support for new approaches and programs that are needed."[52]

For these reasons, the public's right to be represented or consulted during negotiations on the formation of a joint management institution or, at the very least, a right to be heard before an agreement is signed, especially for those who may be personally adversely affected by the agreement, should be recognized. The need to ensure effective public participation in the decision-making processes within the institution implies that all affected groups must be fairly represented among the decision-makers as well. This is especially true with regard to minority groups. Minority groups' interest to be fairly represented at the institutional level derives from the failure of the national political process to allow "discrete and insular minorities" due representation and to exert influence.[53] Minorities are prone to be misrepresented by governments,

---

[51] Lee P. Breckenridge, "Nonprofit Environmental Organizations and the Restructuring of Institutions for Ecosystem Management" (1999) 25 Ecology L.Q. 692 (pointing out that NGOs "constitute a logical place for governmental out-sourcing for technical, resource management, training and other work"). An extensive listing of recent publications addressing the formation in the US of partnerships between governmental and non-governmental groups in the environmental field may be found in Kris Bronars and Sarah Michaels, Annotated Bibliography on Partnerships for Natural Resource Management (1997), available at http://www.icls.harvard.edu/ppp/contents.htm# sources (the website is maintained by the Institute for Cultural Landscape Studies of the Arnold Arboretum, Harvard University).

[52] The International Joint Commission (IJC), *Second Biennial Report under the Great Lakes Water Quality Agreement of 1978 to the Governments of the United States and Canada and the States and Provinces of the Great Lakes Basin 1* (1984), cited in Hayton and Utton, "Transboundary Groundwater," note 40, at p. 710. On the work of the IJC, see, for example, Patricia K. Wouters, "Allocation of the Non-Navigational Uses of International Watercourses: Efforts at Codification and the Experience of Canada and the United States" (1992) 30 Can Yb. Int'l. L. 43.

[53] For this approach, see John Hart Ely, *Democracy and Distrust: A Theory of Judicial Review* (Cambridge, MA, Harvard University Press, 1980), chapter 5. See also Robert M. Cover, "The Origins of Judicial Activism in the Protection of Minorities" (1982) 91 Yale L. J. 1287; Bruce A. Ackerman, "Beyond Carolene Products" (1985) 98 Harv. L. Rev. 713; Owen M. Fiss, "The Supreme Court, 1978 Term – Forward: The Forms of Justice" (1979) 93 Harv. L. Rev. 1.

their perspective lost on decision-makers, and, hence, their interests and values disproportionately affected by the institution's bodies. This is most acute in the case of indigenous groups whose well-being is closely linked to ecosystem management.[54] Procedural guarantees – primarily of a right for fair representation – are, therefore, crucial for promoting their interests.

Note that recognition of the right to participate gives rise to questions related to the definition of who has standing to participate and which modalities of participation would be optimal. Experience, especially in the United States, provides us with a range of examples of participatory options.[55] Another difficulty involved in recognizing this right is that increased importance of NGO participation in transnational institutions would necessitate paying more attention to the identity of the participating NGOs, to prevent possible abuse of their standing by unscrupulous actors.[56] As nonprofit organizations gain influence in the management and allocation of natural resources, taking on functions that are both more governmental and more entrepreneurial, questions of their accountability and fairness are bound to arise.[57] In transnational institutions, the question of who should be granted standing and who should be denied standing would be yet another matter for joint decision.

Control of the agenda

Transnational ecosystem institutions must have the authority to initiate actions. They should have discretion to launch studies on the condition of the shared resource and of the risks of its uses, to conduct inspections of activities that affect or may affect the resource, as well as to embark on long-term planning of future uses. When one party requests its intervention, the institution should have the authority to comply with the request without the consent of the other parties. Controlling its own agenda enables the institution to respond promptly to crises that escape

---

[54] On the perspective of indigenous groups, see chapter 5, notes 9–11 and accompanying text.

[55] See, e.g., John C. Duncan, "Multicultural Participation in the Public Hearing Process: Some Theoretical, Pragmatical and Analytical Considerations," (1999) Colum. J. Envnt'l L. 169; John S. Applegate, "Beyond the Usual Suspects: The Use of Citizens Advisory Boards in Environmental Decisionmaking" (1998) 73 Ind. L.J. 903; Jim Rossi, "Participation Run Amok: The Costs of Mass Participation for Deliberative Agency Decisionmaking" (1997) Nw U.L. Rev. 173.

[56] Breckenridge, "Nonprofit Environmental Organizations," note 51, at p. 698.

[57] Burton A. Weisbrod, "The Future of the Nonprofit Sector: Its Entwining with Private Enterprise and Government" (1997) 16 J. Pol'y Analysis & Mgmt. 541.

the attention of the national governments. More importantly, it reduces the likelihood of inter-governmental collusion to impose unreasonable burdens on politically disenfranchised groups. Moreover, governments often choose to turn a blind eye to other governments' violations not necessarily due to collusion: there are always political costs entailed in initiating rebukes, and therefore governments tend to underproduce them.[58] Empowering the institution to react to violations instead of governments can resolve this prevalent collective-action failure.

The experience with the US–Canada International Joint Commission (IJC) reinforces the imperativeness of such discretion. Under the 1909 treaty, the IJC does not control its own agenda and, hence, must await a reference from both governments. The two governments have often waited too long before referring to the IJC, and the delays have resulted in damage to the environment. This has prompted "widely expressed frustration" within the IJC, which has been prevented from taking timely action against oncoming crises due to its lack of authority to initiate action. This deficiency has motivated calls to empower the IJC to intervene in emerging conflicts on its own initiative.[59]

Review procedures

The question of judicial review becomes crucial when transnational institutions are granted the authority to issue decisions that have binding force on the participating states. We cannot expect such institutions, despite their careful design, to maintain absolute impartiality. Power corrupts, and transnational institutions that are not subject to any external scrutiny will be no exception. Two questions arise. First, what role could judicial review play? More specifically, can adjudicators second-guess institutions' decisions? Second, what type of review process is most appropriate? Specifically, are transnational courts preferable to national ones?

It is my view that judicial review may be beneficial to ensuring the proper functioning of transnational institutions. Although national governments and NGOs can be effective in monitoring the activities of these institutions, their cries of protest may be deemed motivated by self-interest and thus dismissed by other actors as false or illegitimate.

---

[58] Anne-Marie Slaughter, Andrew S. Tulumello, and Stepan Wood, "International Law and International Relations Theory: A New Generation of Interdisciplinary Scholarship" (1998) 92 AJIL 367.

[59] See Hayton and Utton, note 40, at p. 694.

Judicial review could and should emphasize the procedural aspects out-lined in this chapter, rather than second-guess issues of substance.[60] In general, a wide margin of appreciation should be given to the institu-tion's balancing of the different claims and considerations, provided all interests have been properly discussed in due process. Adjudicators are less qualified than the experts and bureaucrats in the institutions to reach an appropriate balancing of competing claims. Yet, they are more qualified to examine whether procedural rules have been adhered to. They also may be more sensitive to procedural shortfalls that hinder the full presentation and weighing of claims of minorities, especially indigenous groups, whom the political process may place at a disadvan-tage and whose interests in the ecosystem are often disregarded. When such groups are affected, adjudicators could prove crucial to ensuring that their interests are properly considered. Therefore, while the margin of appreciation doctrine may theoretically be justified as motivated by the necessity to relegate authority to specialized bodies, a caveat must exist for cases in which minority interests are implicated. A more search-ing judicial inquiry, without recourse to the margin rhetoric, will clear the way for more effective international protection of minority interests in matters concerning the allocation of resources or burdens.[61]

This concern with minority interests also weighs heavily in favor of transnational adjudication as opposed to national judicial review pro-cesses. The gist of the argument is that there are often several groups within every community that tend to be consistently outvoted and, hence, underrepresented in the political process. They are the "discrete and insular minorities" who are, in a very real sense, the political cap-tives of the majority. These groups usually include members of ethnic, national, or religious communities, who are at a numerical disadvantage to the rest of the population.[62] In addition to having different cultures,

---

[60] See also François du Bois, "Social Justice and the Judicial Enforcement of Environmental Rights and Duties" in Alan E. Boyle and Michael R. Anderson (eds.), *Human Rights Approaches to Environmental Protection* (Oxford, Clarendon Press, 1996), pp. 153, 173–4.

[61] See Eyal Benvenisti, "Margin of Appreciation, Consensus and Universal Standards" (1999) 31 NYU J. Int'l L. & Pol. 843.

[62] Compare Capotorti's widely accepted definition of minorities as "a group numerically inferior to the rest of the population of a State, in a non-dominant position, whose members – being nationals of the State – possess ethnic, religious or linguistic characteristics differing from those of the rest of the population and show, if only implicitly, a sense of solidarity, directed towards preserving their culture, traditions, religion or language." Francesco Capotorti, *Study on the Rights of Persons Belonging to Ethnic, Religious and Linguistic Minorities* (New York, United Nations, 1979) E/CN.4/Sub.2/384/Rev.1, at p. 96.

traditions, and, sometimes, appearances from the majority, the loyalty of these groups to the majority-controlled institutions is often called into question by members of the majority, and wariness with regard to potential irredentism and secessionism is rife. Absent political influence and faced with widespread resentment, minorities rely upon the judicial process to secure their interests.[63] But because the national judicial process – in itself dominated by judges belonging to the majority – may fail to protect them, international judicial and monitoring organs are often their only reliable and final resort. In conflicts related to ecosystem management, which often have outcomes that exclusively or predominantly place a burden on the rights and interests of minorities, no preference for national adjudication is warranted. In such conflicts, supranational institutions that are staffed not only by representatives of governments are preferable for ensuring the rights of nationally underrepresented groups. Good precedents for this point have been set by international human rights bodies that were able to safeguard minority interests also with respect to the allocation of resources between the minority and majority groups. National plans to reduce, for example, grazing areas crucial for maintaining the culture of the Saami minority in Finland[64] were scrutinized strictly by the Human Rights Committee, which refused to defer to the state's margin of appreciation.

Other considerations that support the preference for a transnational, rather than national, review process emphasize aspects of efficiency. Decisions by a transnational body would be final and binding upon both the institution and the participating states. The panel of adjudicators would include also representatives with expertise in the specific matter at issue. The above-mentioned concern for minority rights requires that their representatives also be included on the panel of adjudicators.

Institution formation and reformation

As discussed in chapter 4, the setting-up of transnational institutions is in itself a collective-action problem and can entail attempts to capture opportunistic gains. The process of designing such institutions should allow for the participation of the wider public, both through

---

[63] See Ely, *Democracy and Distrust*, note 53.

[64] See the view of the Human Rights Committee under the Optional Protocol of the 1966 International Covenant on Civil and Political Rights (ICCPR) (finding that reindeer husbandry is an essential element of Saami culture and, as such, protected under Article 27 of the Covenant): *Länsman et al.* v. *Finland*, Communication No. 511/1992, UN Doc. CCPR/C/52/D/511/1992 (1994).

representative NGOs and through the dissemination of accessible information. This designing process is not an easy task. A delicate balance must be found to accommodate governmental, inter-governmental, and non-governmental representation and to ensure that narrow interests, including those advanced by NGOs, do not gain dominance.

The establishment of institutions could follow a transformational pattern, namely, a modest start that evolves incrementally towards thick and binding cooperation. Advocates of transformationalism suggest that the initial iteration of relatively low-cost interstate commitments tends to enhance reliance and increase the willingness to commit further.[65] This theory is supported by the experience with existing regimes, although there is also evidence that a different approach, one that encompasses fewer parties but provides for deep initial cooperation, could yield better results.[66] This debate – whether to prefer inclusion to exclusion, shallow to deep cooperation – may well be moot in the transnational ecosystem context. This is because it makes little sense to exclude parties who hold stakes in the shared resource. In addition, transnational management will not have any significance unless deep cooperation is ensured from the initial stage. In any event, whatever the level of cooperation achieved at the initial stage, parties should allow for the modification of the institution they have established. As the International Court of Justice recognized in the *Gabcikovo–Nagymaros* decision, flexibility is a prerequisite for long-term cooperation. Transnational ecosystem institutions must remain flexible enough to enable modifications that respond to changed circumstances, to new information that reveals errors in the structural design, and to enhanced willingness to cooperate on the part of the member states. Any joint-management mechanism must provide rules and procedures for its own modification. For the same reasons discussed earlier, this must not be left to representatives of governments negotiating behind closed doors. Rather, the procedure must include also scientific experts, representatives of minority groups, and NGOs representing diverse interests. With the uses and allocations of the ecosystem shares under constant reappraisal,

---

[65] See, e.g., Jutta Brunnée and Stephen J. Toope, "Environmental Security and Freshwater Resources: Ecosystem Regime Building" (1997) 91 AJIL 26, 46; Chayes and Chayes, *New Sovereignty*, note 46; Marc A. Levy *et al.*, "Improving the Effectiveness of International Environmental Institutions" in Peter M. Haas, Robert O. Keohane, and Marc A. Levy (eds.), *Institutions for the Earth: Sources of Effective International Environmental Protection*, Global Environmental Accords Series (Cambridge, MA, MIT Press, 1993), pp. 397, 413.

[66] George W. Downs *et al.*, "The Transformational Model of International Regime Design: Triumph of Hope or Experience?" (2000) 30 Colum. J. Trans. L. 465.

renegotiations are channeled to the treaty bodies, away from potentially divisive domestic forums. Although there is always the risk that efforts to renegotiate problematic agreements will fail, it is significantly lower than the risks presented by unilateral abuse of any of the diverse "escape doctrines" from treaty obligations.

Conclusion

This chapter outlines the modalities for the operation of transnational ecosystem management institutions. I have suggested that institutions that follow the principles identified in this chapter are likely to adopt policies that are efficient, equitable, and sustainable and reflect a fair and balanced weighing of the interests of all communities sharing the common resource or resources. They are likely to perform well the functions identified in the previous chapter. The task of the next chapter will be to examine to what extent these principles are already reflected in current international law and what still remains to be recognized.

7 The development of positive international law on transboundary ecosystems: a critical analysis

Law at the crossroads of two clashing philosophies

This chapter examines the arrangements international law can offer or develop in response to the challenges identified in the previous chapters. Thus far, we have investigated and identified the goals and constraints involved in the management of transnational ecosystems and the causes of collective-action failure that all too often hinder states from achieving these goals. We also have considered the optimal remedies to such failures. The previous chapter delineated the necessary elements for the optimal and sustainable operation of collective institutions. The question that remains is whether states can agree to accept these basic elements and act collectively through institutions. At this juncture, we turn to international law, whose task is to provide proper incentives – both positive and negative – for states to cooperate through such institutions. In this chapter, I examine the positive law that deals with transboundary ecosystems and inquire as to what extent it reflects sound policies and to what extent – and how – it could be modified to provide the proper incentives. True to this book's confidence in the idea of endogenously evolving collective action, this chapter argues that international law can create incentives for states to cooperate. It can do so by providing states with incentives to enter into negotiations (rather than opt for adjudication of outstanding disputes) and by strengthening the institutions they may eventually form.

It is quite unorthodox for a legal treatise whose main focus is the analysis of specific norms, particularly norms of international law, to defer discussion of the positive law to such a late stage. But in a sphere that is fraught with disagreements over basic norms, nurtured by conflicting philosophies as to the proper approach to be taken, this is arguably the best way to distinguish between appropriate norms and inappropriate ones, as well as to highlight existing discrepancies between the positive law and its alternatives.

As we trace the evolution of the relevant norms of international law, we will notice a clash between two philosophies. The first philosophy – what I call "the philosophy of disengagement" – characterizes the attitude of many governments as well as of the World Bank, whereas the second – "the philosophy of integration" – characterizes the approach advocated mainly by scientists and scholars, but particularly by judges in international litigation. The two contradictory philosophies pushed the codification efforts of international ecosystem law throughout the twentieth century in two opposite directions. Whereas the disengagement philosophy strives to limit common ownership among riparian states to the lowest possible minimum, the philosophy of integration suggests that common ownership and inclusive management is not only an inescapable outcome, but also a beneficial one. The disengagers look to international law as a system of rules that could minimize friction among riparians and resolve interstate disputes through adjudication as well as rigid arm's-length agreements assigning allocation of rights and obligations as clearly as possible. In contrast, the integratives seek more, rather than less, friction as the preferred alternative and, hence, opt for the management of disputes through negotiations leading to flexible agreements that establish joint management institutions. The integratives acknowledge that states are no longer "bound by international law only at their own discretion [but instead have] the responsibility to develop and implement international law in order to further the interests of humankind."[1] These differences are reflected in the definition of shares of international resources, whether subject to unilateral or common ownership; in the different character and role given to norms defining entitlements and prescribing procedure; and in the different modalities for dispute resolution. Oddly enough, 1997 witnessed the climax of both philosophies. The philosophy of disengagement won out in the 1997 UN Convention on the Non-Navigational Uses of International Watercourses ("the Watercourses Convention");[2]

---

[1] Ellen Hey, "Sustainable Use of Shared Water Resources: The Need for a Paradigmatic Shift in International Watercourses Law" in Gerald H. Blake *et al.* (ed.), *The Peaceful Management of Transboundary Resources* (London, Graham & Trotman, 1995), p. 127.

[2] United Nations Convention on the Law of the Non-Navigational Uses of International Watercourses (adopted on 21 May 1997), reprinted in (1997) 36 ILM 700.

the philosophy of integration prevailed four months later in the ICJ decision in the *Gabcikovo-Nagymaros* case.[3]

The next part describes the norms designed to promote negotiation among states sharing a common ecosystem. The subsequent part discusses the norms that strengthen institutions, in particular, transnational ecosystem institutions.

Negotiation-enhancing norms

For proponents of the integrative philosophy – namely, the establishment of transnational ecosystem institutions for the management of shared resources – the preferred legal environment is one that promotes negotiation rather than litigation. As noted, most disputes concerning shared water resources have been resolved by negotiation and not litigation.[4] The satisfactory resolution of water-related disputes through litigation in a handful of cases[5] cannot conceal the fact that this mode of action suffers from fundamental deficiencies. "Water disputes are generally agreed to constitute a classical example of disputes which cannot be satisfactorily solved by judicial decision."[6] Other environmental disputes have fared no better, due to a number of factors. The first factor is the prevalent governmental reluctance to relinquish control over the fate of strategic resources in submitting to adjudication. In contrast, not only do negotiations ensure complete control of the outcome, they also present a real assurance of efficient and equitable outcomes. In international settings, negotiations, as opposed to arbitration and adjudication, are the key to initiating cooperation, promoting confidence between riparians, and increasing the range of channels of communication and cooperation, even in spheres unrelated to the resource. Indeed, even in

---

[3] Gabcikovo–Nagymaros Project (Hungary/Slovakia), Judgment, ICJ Reports 1997, p. 7, reprinted in http://www.icj-cij.org/idocket/ihs/ihsjudgement/ihsjudframe1.htm; (1998) 37 ILM 167.

[4] See, e.g., Charles B. Bourne, "Mediation, Conciliation and Adjudication in the Settlement of International Drainage Basin Disputes" (1971) 9 Can. Yb. Int'l L. 114.

[5] On these cases, see Stephen S. McCaffrey, "Second Report on the Law of the Non-Navigational Uses of International Watercourses" Doc. A/CN.4/399 (1986) 2 *ILC Yearbook* (Part 1) 87, 113–22; "Legal Problems Relating to the Utilization and Use of International Rivers" Doc. A/5409 (1974) 2 *ILC Yearbook* (Part 2) 33, 187–99.

[6] Friedrich J. Berber, *Rivers in International Law* (London, Stevens, 1959), p. 263. See also Jerome Lipper, "Equitable Utilization" in Albert H. Garretson, Robert D. Hayton, and Cecil J. Olmstead (eds.), *The Law of International Drainage Basins* (Dobbs Ferry, NY, Oceana Publications, 1967), pp. 15, 59–60.

federal systems where courts can be quite effective, we find a preference for interstate negotiation over litigation.[7] Therefore, to promote cooperation, international law should provide states with ample incentives for negotiating.

The benefits of direct negotiations over litigation as a means of settling international water disputes are underscored in the growing analytical literature on negotiations, in general, and on international negotiations, in particular. This literature draws upon game theory and psychology, pointing to gains that extend even beyond the resolved dispute.[8] For our purposes, two crucial benefits stemming from international negotiations are important to consider. The first benefit is the increased likelihood of reaching an efficient result. Negotiations provide the parties with an opportunity to exchange information about their domestic constraints and explore their differences in valuation, preference, risk aversion, and in other dimensions.[9] The argument, well supported

[7] In the United States, the preferred tool is interstate compacts: see Joseph L. Sax, Robert H. Abrams, and Barton H. Thompson, Jr., *Legal Control of Water Resources: Cases and Materials* (2nd edn, St. Paul, MN, West Publishing Company, 1991), pp. 733–46. In India, interstate disputes are negotiated with the aid of the central administration. S. N. Jain, Alice Jacob, and Subhash C. Jain, *Interstate Water Disputes in India: Suggestions for Reform in Law* (Bombay, NM Tripathi, 1971) (especially chapter 2); N. D. Gulhati, *Development of Inter-State Rivers: Law and Practice of India* (Bombay, Allied Publ., 1972) pp. 53–6. Note that courts in federal states have, at times, allocated water shares to states. For the situation in the US, see *Arizona* v. *California*, 373 US 546 (1963); *Colorado* v. *New Mexico*, 459 US 176, 103 S. Ct. 539, 74 L.Ed.2d. 348 (1982). In Argentina, see *Province of La Pampa* v. *Province of Mendoza*, Argentina Supreme Court of Justice, December 1987 (reported in *International Rivers and Lakes* No. 10, May 1988, at 2). In Germany, see *Wuerttemberg & Prussia* v. *Baden* (the *Donauversinkung* case), German Staatsgerichthof, 18 June 1927, reprinted in A. MacNair and H. Lauterpacht (eds.), *Annual Digest of Public International Law Cases* (London, Longmans, Green & Co., 1927-8), p. 128 vol. IV. In Switzerland, see *Aargau* v. *Zurich, Schweizerische Bundesgericht*, 1878 (cited in William L. Griffin, "The Use of Waters of International Drainage Basins under Customary International Law" (1959) 53 AJIL 50, 66).

[8] There is extensive literature in this area. Among the leading treatises are Roger Fisher and William Ury, *Getting to Yes: Negotiating Agreement without Giving In* (Boston, Houghton Mifflin, 1981); Oran R. Young, *Bargaining: Formal Theories of Negotiation* (Urban, University of Illinois Press, 1975); Howard Raiffa, *The Art and Science of Negotiation* (Cambridge, MA, Belknap Press of Harvard University Press, 1982); David A. Lax and James K. Sebenius, *The Manager as Negotiator* (Washington, DC, National Institute for Dispute Resolution, 1985). On international negotiations, see Raiffa, *Art and Science*; James K. Sebenius, *Negotiating the Law of the Sea* (Cambridge, MA, Harvard University Press, 1984); Victor A. Kremenyuk (ed.), *International Negotiations: Analysis, Approaches, Issues* (San Francisco, Jossey-Bass Publishers, 1991).

[9] See Sebenius, *Negotiating*, note 8, at p. 114. On the means for exploring the differences, see pp. 117–44.

by the literature on negotiations theory, is that the parties are in a better position than third parties such as judges or arbitrators to do this and therefore reach an efficient and stable outcome.[10] The second important benefit of direct negotiations is the potential establishment of long-term relationships based on mutual respect. Negotiations, both formal and informal, direct and indirect, and, in particular, the exchange of information on the parties' needs and constraints can lead to the fading of the negative image of the other side and the removal of cultural barriers that so often impede cooperation.[11] The adoption of a more perceptive and considerate appreciation of the constraints of the other parties increases significantly the likelihood of establishing on-going and lasting relationships beyond the signing of the agreement. In negotiations, the parties embark on an incremental process of building up relationships that may culminate in multi-level cooperation. In contrast to a third party settlement, which leaves the parties in an adversarial frame of mind that is not conducive to further cooperation, the negotiation experience creates an impetus for formalizing the interaction through the establishment of institutions for joint management of the shared transboundary resources. Litigation cannot impose upon litigants long-term cooperation; any cooperation through a joint institution can only be the product of negotiation. When the negotiating parties' interaction encompasses other spheres in addition to the specific shared resource, the likelihood of stronger ties increases. As noted by collective-action theorists, cooperation in the use of a common pool resource is most likely to be reinforced by cooperation in other spheres as well.[12] Negotiating parties – as opposed to third party decision-makers – often enhance the agreement zone through the addition of other contentious issues, thus securing sustainable "package deals."[13] Multidimensional deals, such as the Israeli–Palestinian and Israeli–Jordanian agreements, are examples of "thick" interdependence

[10] See Lax and Sebenius, *The Manager*, note 8, ch. 5; Arild Underdal, "The Outcomes of Negotiation" in *International Negotiations*, note 8, pp. 100–15; James K. Sebenius, "Negotiation Analysis" in *International Negotiations*, note 8, at pp. 203, 210.

[11] On cultural barriers in the context of water disputes, see Guy Olivier Faure and Jeffrey Z. Rubin (eds.), *Culture and Negotiation: The Resolution of Water Disputes* (Newbury Park, CA, Sage Publications, 1993).

[12] On multidimensional relations as reinforcing cooperation, see Elinor Ostrom, *Governing the Commons: The Evolution of Institutions for Collective Action* (Cambridge, Cambridge University Press, 1990), p. 207; Russell Hardin, *Collective Action* (Baltimore, Johns Hopkins University Press, 1982), pp. 31–3.

[13] Sebenius, *Negotiating*, note 8, at p. 198. This is true unless the other contentious issues are more difficult politically; *ibid.* at p. 200.

in which one of the crucial topics is the utilization of shared water resources.[14]

International law can persuade parties to negotiate with a view to co-operation rather than resorting to adversarial litigation. It can, and does, require riparians to negotiate in good faith towards an agreement.[15] However, dictating such a requirement is not enough. My claim is that the law, by prescribing a number of background norms, can effectively induce negotiations between riparians. These norms are discussed in the next section below.

Definition of allocation criteria: vague standards versus clear rules

The background norms that assign shares to states sharing a resource can shape the choice of those states as to whether to pursue negotiations or adjudication. The argument here is that a vague standard impels riparians to seek a negotiated agreement rather than litigate towards an unpredictable result.[16] This is because the vague standard leaves enough leeway for negotiations, during which each side can develop its own interpretation of the standard, whereas clearer rules

---

[14] See Israel–PLO Declaration of Principles on Interim Self-Government Arrangements, 13 Sept. 1993, reprinted in (1993) 32 ILM 1525, Annex II; and the Israeli–Palestinian Interim Agreement on the West Bank and the Gaza Strip, Washington, 28 September 1995, Annex III, Protocol Concerning Civil Affairs, Appendix 1 – Powers and Responsibilities for Civil Affairs, Article 40 (Water and Sewage), and Schedule B; Israel–Jordan Treaty of Peace of 26 October 1994, (1995) 324 ILM 46 (Annex II).

[15] See Articles 3 and 6 of the Institute of International Law's Resolution on the Utilization of Non-Maritime International Waters (Except for Navigation), adopted at its session at Salzburg (3-12 September 1961) (1961) 49 (II) Ann. Inst. Dr. Int'l 370 (translated in 56 AJIL 737 [1962]; Article 6 of the 1966 Helsinki Rules (ILA Report of the Fifty-Second Conference 484 (1967)). But see a very nuanced duty under the Watercourses Convention, note 2, Article 3(5), which stipulates that when one party seeks the adjustment and application of the Convention, "States shall consult with the view to negotiating in good faith." On the duty to negotiate in good faith, see, for example, Julio A. Barberis, "Bilan de recherches de la section de langue française du Centre d'étude et de recherche de l'Académie" in Centre for Studies and Research, *Rights and Duties of Riparian States of International Rivers* (Dordrecht, M. Nijhoff, 1990), pp. 15, 54–5; Janos Bruhacs, *The Law of Non-Navigational Uses of International Watercourses* (Dordrecht, M. Nijhoff, 1993), pp. 176–8; Charles B. Bourne, "Procedure in the Development of International Drainage Basins: The Duty to Consult and to Negotiate" (1972) 10 Can. Yb. Int'l L. 212, 224–33; Dominique Alhértière, "Settlement of Public International Disputes on Shared Resources: Elements of a Comparative Study of International Instruments" (1985) 25 Nat. Res. J. 701.

[16] This point is mentioned by David Caron, "The Frog that Wouldn't Leap: The International Law Commission and its Work on International Watercourses" (1992) 3 Colo. J. Int'l Envt'l L. & Pol'y 269, 273.

prevent many riparians – clear winners or clear losers – from entering the negotiations room. Negotiations under a vague standard would involve not only lawyers and politicians who are concerned with the domestic political ramifications of the "concessions" offered to neighboring states; they would also include scientists from various disciplines who could suggest alternatives for optimal solutions and whose contributions could defuse the potentially adversarial tone of the negotiations.[17] With the shared language of technical expertise, political constraints can be sidestepped and decisions based on objective data more easily reached.[18]

Ambiguity in the law often assists parties who seek to defer or avoid stumbling blocks when negotiating an agreement.[19] It enables them to avoid troublesome issues by, for example, assigning ad hoc committees to further explore the possibility of a negotiated agreement on particular matters. Such committees are often the precursors to permanent transnational institutions. Thus, for example, the vague standard of "equitable utilization" proved useful during the negotiation of the Israeli–Palestinian Declaration of Principles of 1993, which deferred the potentially explosive issue of water allocation to a later stage in the discussions.[20] According to Annex III of the Declaration, cooperation on water is the first topic on the agenda of the joint Continuing Committee for Economic Cooperation. Such cooperation was to begin with an examination of "proposals for studies and plans on water rights of each party, as well as on the equitable utilization of joint water resources for implementation in and beyond the interim period."[21] If the law on apportionment of water resources had been any clearer, the costs of tackling (and deferring) the water issue in the Israeli–Palestinian context would have been very high, if not prohibitive. The existence of this vague standard also enabled Israel and Jordan to agree, already in September 1993, on a Common Agenda for the Bilateral Peace Negotiations, which was to

---

[17] Such early reference is recommended in the Bellagio Draft Treaty, a suggested blueprint for the joint management of aquifers. See Robert D. Hayton and Albert E. Utton, "Transboundary Groundwater: The Bellagio Draft Treaty" (1989) 29 Nat. Res. J. 663, 714–16.

[18] One example of such a technical internal procedure can be found in the Namibia–South Africa Agreement on the Establishment of a Permanent Water Commission (14 Sept. 1992), reprinted in (1993) 32 ILM 1147. The Agreement establishes a joint commission to serve as technical adviser to the state parties. Its task, among other things, is to gather data and advise the parties on "the criteria to be adopted in the allocation and utilization of common water resources."

[19] Sebenius, *Negotiating*, note 8, at p. 125.

[20] See Israel–PLO Declaration of Principles, *supra*, note 14, Annex II.

[21] *Ibid.*, Annex III, section 1.

include the allocation of the Jordan and Yarmuk Rivers, despite serious disagreements that existed at that stage about the outcome of the negotiations. In fact, as though the existing standard was not vague enough, the parties agreed on an even vaguer formula for "securing the rightful water shares of the two sides."[22] The improved atmosphere after the 1994 Washington summit led to continued negotiations and finally – as the last topic to be settled – to a satisfactory solution.[23]

Which kind of norm reflects positive international law? Does the law support a vague standard or a clear rule? There is no easy answer to this seemingly simple question. In fact, this question spawned an extremely divisive debate throughout the International Law Commission's (ILC) work in the area of fresh water, as well as during the negotiations for the 1997 Watercourses Convention. This debate continued up until the very last moment of the work of the Sixth Committee that elaborated the Convention, and almost led to its failure.[24] During these negotiations, there was a clash between the proponents of a vague norm and those favoring a clear norm. The proposed vague standard prescribed allocation of entitlements based on what in a specific basin would be "reasonable and equitable" allocation. Basically, this standard called for the balancing of the needs of the communities that share the common resource, "taking into account all relevant factors and circumstances" in the specific basin, including its natural characteristics as well as the tension between existing and potential uses.[25] This standard competed against the much clearer rule of "no significant harm," which proscribed any "significant" interference with the quality and quantity of the flow from one state to another, "significant" in the sense of "being capable of being established by objective evidence and not trivial in nature, it need not rise to the level of being substantial."[26] Whereas the "equitable and

[22] See section 3 (a) of the Common Agenda. The document is reprinted in (1993) 32 ILM 1522, 1523.

[23] Israel–Jordan Peace Treaty, note 14, Article 6 and Annex II. The water issue remains the last – and most difficult – item on the agenda.

[24] See the Summary Record of the first part of the 62nd meeting of the Sixth Committee A/C.6/51/SR. 62 (1997). For a "behind the scenes" account, see Lucius Caflisch, "Regulation of the Uses of International Watercourses" in Salman M. A. Salman and Laurence Boisson de Chazournes (eds.), *International Watercourses: Enhancing Cooperation and Managing Conflict* (Washington, DC, World Bank, 1998), pp. 3, 13–16.

[25] Article 5 of the Helsinki Rules, *supra*, note 15; Article 6 of Watercourses Convention, note 2.

[26] See the Statements of Understanding pertaining to certain Articles of the Watercourses Convention, the understanding relating to Article 3 (b) (see note 2, at p. 719). The ILC previously had referred to the rule of "no appreciable harm."

reasonable" standard would invite riparians of a specific basin to negotiate what this concept denotes for their resource, the "no harm" rule
would assign rights protected by a "property rule" to riparians' existing
shares.[27] It would provide for a clear "exit option" for each side, i.e., an
alternative to negotiation.[28] Armed with the clear "no harm" rule, riparians – both upstream and downstream – whose current uses satisfy their
needs would have no incentive to negotiate a reduction of their uses,
even if they are wasteful or polluting. Moreover, such a rule runs contrary to the principle of sustainability[29] and the obligation to abate harm
through gradual regulation and elimination of certain pollutants.[30]

See former Draft Article 7, reprinted in (1991) 30 ILM 1575. On this "codification
battle," see Charles B. Bourne, "The Right to Utilize the Waters of International Rivers"
(1965) 3 Can. Yb. Int'l L. 187, 209–10; Charles B. Bourne, "The International Law
Commission's Draft Articles on the Law of International Watercourses: Principles and
Planned Measures" (1992) 3 Colo. J. Int'l Envt'l L. & Pol'y 65, 80–5; Lucius Caflisch,
"Sic utere tuo ut alienum non laedas: Règle prioritaire ou élément servant à mesurer
le droit de participation équitable et raisonnable à l'utilisation d'un cours d'eau
international" in Alexander von Ziegler and Thomas Burckhardt (eds.), *Internationales
Recht auf See und Binnengewässern* (1993), pp. 27, 41–7; Patricia K. Wouters, "Allocation
of the Non-Navigational Uses of International Watercourses: Efforts at Codification and
the Experience of Canada and the United States" (1992) 30 Can. Yb. Int'l L. 43, 80–6.

[27] On the "property rule" and other rules that protect entitlements, see Guido Calabresi
and A. Douglas Melamed, "Property Rules, Liability Rules, and Inalienability: One
View of the Cathedral" (1972) 85 Harv. L. Rev. 1089, 1092: "An entitlement is protected
by a property rule to the extent that someone who wishes to remove the entitlement
from its holder must buy it from him in a voluntary transaction in which the value of
the entitlement is agreed upon by the seller."

[28] On legal rules as exit options, see Douglas G. Baird, Robert H. Gertner, and Randal
C. Picker, *Game Theory and the Law* (Cambridge, MA, Harvard University Press, 1994),
pp. 224–32. Exit options are the legal rights available to actors when no consensual
bargain is reached.

[29] On inter-generational equity, the moral underpinning of the sustainable use principle,
see chapter 5, text accompanying notes 45–56.

[30] See, for example, the following specific and general instruments: United
States–Canada Agreement on Great Lakes Water Quality, (1978) 30 UST 1383, TIAS
9257; Convention on the Protection of the Rhine Against Chemical Pollution, (1976)
1124 UNTS 375, (1977) 16 ILM 242; Convention for the Protection of the Rhine from
Pollution by Chloride, 1976, reprinted in (1977) 16 ILM 265; 1992 Rio Declaration on
Environment and Development, Principles 7 & 8 (reprinted in [1992] 31 ILM 874; ILA's
Montreal Resolution, Article 1[c] (states shall attempt to reduce water pollution to
the lowest "practicable and reasonable" level) (International Law Association, Report of
the Sixtieth Conference, Montreal, 1982); Institut de droit international's Athens
Resolution on the Pollution of Rivers and Lakes and International Law, Article 3 (1)(b)
(states shall abate existing pollution "dans les meilleurs delais") ([1980] 58(II) Ann.
Inst. Dr. Int'l 1979, 198–99); ECE Declaration of Policy on Prevention and Control of
Water Pollution, Including Transboundary Pollution (Decision B[XXXV] 1980)
(reprinted in Economic Commission for Europe, *Two Decades of Cooperation*

The choice between a vague standard and a clear rule has important implications beyond the negotiations phase as well. The vague standard for current and future allocations and reallocations enables the agreement to be updated in a sufficiently flexible way to meet changes in supply and demand. Thus, this standard enhances cooperation beyond the conclusion of the agreement on initial shares.[31]

The proponents of the "no harm" rule comprised two camps. The first camp was riparians who subscribed to the disengagement philosophy and had no interest in cooperation. These were the relatively stronger riparians, the regional bullies, who had managed, over the years, to ensure themselves a constant supply of the resource and now wanted to protect their shares against any "harm" done by their upstream neighbors and any new demands by lower riparians. The "no harm" rule was preferred by downstream states like Egypt, who resisted any diminution of the Nile waters, as well as by upstream China and Turkey, who refused to agree to any implied restraints on their established uses of (respectively) the Mekong and the Euphrates.[32]

The other active voice in favor of the "no harm" rule was the World Bank, which sought to facilitate unilateral investments in water-related infrastructure and was instrumental in mounting resistance to the integrative approach. Since 1948, the World Bank has lent over US\$34 billion (13 percent of total lending) for water development.[33] In making these loans, the World Bank has committed itself to "act[ing] prudently in the interests both of the particular member in whose territories the

on Water [1988]) at p. 1: Principle 4 (governments should adopt measures reducing existing water pollution). Compare 1966 Helsinki Rules, note 15, Article X(1) (b) (states "should take all reasonable measures to abate existing water pollution in an international drainage basin *to such an extent that no substantial damage is caused in the territory of a co-basin State*" [emphasis added]).

[31] On the flexibility of long-term agreements concerning transboundary resources and fresh water in particular, see discussion at chapter 5, notes 81–2 and accompanying text.

[32] See the statements of the delegates of Turkey, Egypt, and China in Summary Record of the First Part of the 62nd Meeting (A/C.6/51/SR.62), p. 2 (Turkey), and Summary Record of the Second Part of the 62nd Meeting (A/C.6/51/SR.62/Add. 1), pp. 3–4 (Egypt), p. 5 (China).

[33] See Raj Krishna, "The Evolution and Context of the Bank Policy for Projects on International Waterways" in Salman and Boisson de Chazournes, *International Watercourses*, note 24, p. 31; Raj Krishna, "International Watercourses: World Bank Experience and Policy" in J. A. Allan and Chibli Mallat (eds.), *Water in the Middle East: Legal, Political and Commercial Implications* (London, I. B. Tauris Publishers, 1995), p. 29.

project is located and of the members as a whole."[34] The Bank, therefore, is bound to comply with international law so as to avoid harm to other riparians. In 1949, the Bank began to require the consent of all riparians as a precondition to financing unilateral projects on international rivers. Soon enough, however, the political implications of this policy struck: Egypt, which had been refused financing of its proposed High Aswan Dam project, turned to the Soviet Union for assistance. A revision of the Bank's policy soon followed. In 1956, it relaxed its consent requirement and, instead, required only an examination of whether a proposed project would be "harmful to the interests of the other riparians."[35] The "no harm" rule was the test used in examining such proposals. The alternative assessment, based on the standard of equitable and reasonable share, was rejected because it would render the process of approval of unilateral plans extremely complex and would require the cooperation of all riparians,[36] thus, in effect, again subjecting proposals to the consent of all riparian neighbors. This was made explicit in the Bank's 1985 revised policy, in which the "no appreciable harm" rule became "the building block" of the Bank's water development policy.[37]

The "integrative" camp of the "equitable and reasonable share" consisted of riparians who had committed themselves to cooperation or were otherwise eager to use international law to support their efforts to modify existing allocations. They relied on the emerging broad recognition of the "equitable use" doctrine, which, by then, had been recognized by the Institut de droit international (IDI), the International Law Association (ILA), and the Economic Commission for Europe (ECE), as reflecting customary international law.[38]

Initially, the "no harm" rule enjoyed the support of the majority at the ILC, whose Draft Article 7, provisionally adopted on first reading in

---

[34] Article III 4(v) of the Bank's Articles of Agreement (appears in Krishna, "International Watercourses", note 33, at p. 31).

[35] Operational Statement No. 5.05 (appears in Krishna, "International Watercourses", note 33, at 35.

[36] *Ibid.*, at p. 45.   [37] *Ibid.*, at p. 44.

[38] The ECE-sponsored convention was the 1992 Helsinki Convention on the Protection and Use of Transboundary Watercourses and International Lakes, reprinted in (1992) 31 ILM 1312 ("The Helsinki Convention") (see Article 2[2][c]). For an impressive number of regional treaties that adopted the same principle either expressly or implicitly, see Barberis, "Bilan de recherches," note 15, pp. 38–47. The same principle was applied as early as 1906 to the apportionment of the waters of the Rio Grande River between Mexico and the US: Neal. E Armstrong, "Anticipatory Transboundary Water Needs and Issues in the Mexico–US Border Region in the Rio Grande Basin" (1982) 22 Nat. Res. J. 877, 904.

1991, gave unambiguous precedence to the duty to prevent "appreciable harm," stating that "watercourse States shall utilize an international watercourse in such a way as not to cause appreciable harm to other watercourse States."[39] In round two, at the second and final reading, the equitable use standard prevailed, as acknowledged in section 2 of a newly formulated Draft Article 7:

where, despite the exercise of due diligence, significant harm is caused to another watercourse State, the State whose use causes the harm shall, in the absence of agreement to such use, consult with the State suffering such harm over:

   (i)   The extent to which such use is equitable and reasonable taking into account the factors listed in article 6;

   (ii) The question of ad-hoc adjustments to its utilization, designed to eliminate or mitigate any such harm caused and, where appropriate, the question of compensation.[40]

Ultimately, during the discussions of the Sixth Committee, a real effort was made to square the circle and strike a balance between the two irreconcilable approaches in order to gain sufficient votes for the adoption of the Watercourses Convention. This, in fact, was an effort to blur the distinction so that both camps would be able claim victory. The final Article 7(2) reads:

(2) Where significant harm is nevertheless caused to another watercourse State, the State whose use causes the harm shall, in the absence of agreement to such use, take all appropriate measures, having due regard for the provision of articles 5 and 6, in consultation with the affected State, to eliminate or mitigate such harm and, where appropriate, to discuss the question of compensation.[41]

The omission of the "due diligence" caveat, the inclusion of the clearer duty to "take all appropriate measures," and, ultimately, the replacement of "taking into account" with "having due regard" were deemed by the proponents of the "no harm" camp sufficient to indicate that the "no harm" duty had prevailed, while at the same time, the "equitable use" camp could also celebrate success.

Be that as it may, the dust had barely settled on this legal front when the ICJ stepped into the picture with its *Gabcikovo–Nagymaros* decision.[42] The decision cut the Gordian knot of the convoluted norm by referring

---

[39] Reprinted in (1991) 30 ILM 1575.
[40] For the text of the draft articles and commentary, see *Report of the ILC on the Work of its Forty-Sixth Session*, UN GAOR 49th Session Supp. No. 10 (A/49/10) (1994), at 195–326.
[41] The Watercourses Convention, note 2.   [42] See note 3.

only to the "equitable use" concept, crowning it as the true reflection of customary international law.[43] The court did not even mention the "no harm" rule, despite the fact that it would have reinforced the conclusion that Hungary, as the downstream state, was protected against Czechoslovakia's unilateral diversion project. This is one of the numerous manifestations of the clear judicial predilection for the philosophy of integration, as suggested earlier, as opposed to the disengagement philosophy reflected in the Watercourses Convention. This explicit judicial adoption of a cooperation-friendly vague standard and rejection of the litigation-prone disengagement rule demonstrates the utility of judge-made law and its role in developing customary international law, a topic further elaborated on in chapter 8.

Needs versus nature in the analysis of entitlements

The vague standard for allocating shares of a transboundary resource – "equitable and reasonable use" – seems to stir a jumbled mixture of factors into the hot cauldron of international negotiations. Almost everything appears to have relevance: the natural characteristics of the specific basin; the needs of the relevant populations; past and potential uses; and so on and so forth. The Watercourses Convention explicitly rejects any attempt to prioritize the factors, declaring, "In determining what is a reasonable and equitable use, all relevant factors are to be considered together and a conclusion reached on the basis of the whole," while "the weight to be given to each factor [including those factors listed in Article 6(1)] is to be determined by its importance in comparison with that of other relevant factors."[44] This outcome has led to the cautionary warning that "in view of the complexity in the balancing process involved, recourse to the doctrine of equitable use by the interested parties themselves is unlikely to lead to a solution of a conflict that has arisen exactly because of divergent interpretations of what constitutes 'reasonable use' in the circumstances of the case concerned."[45]

---

[43] *Gabcikovo–Nagymaros* case, para. 85: "The court considers that Czechoslovakia, by unilaterally assuming control of a shared resource, and thereby depriving Hungary of its right to an equitable and reasonable share of the natural resources of the Danube . . . failed to respect the proportionality which is required by international law."

[44] Article 6(3) of the Watercourses Convention, note 2.

[45] Gunther Handl, "Balancing of Interests and International Liability for the Pollution of International Watercourses: Customary Principles of Law Revisited" (1975) Can. Yb. Int'l L. 156, 189.

If we take this seemingly open-ended balancing process at face value, this warning is sound. However, the study of state practice does support the argument that customary international law assigns different relative weight to each of the factors on this list. What is noteworthy about this assignment is that the natural conditions – for example, how much rain each state contributes to a river – have a lesser weight in the balancing process. Primary attention is assigned to the needs of the populations dependent on the resource. This was noted by the IDI's 1961 Resolution, which suggested that settlement of disputes over utilization between states "will take care on the basis of equity, taking particular account of their respective needs, as well as other pertinent circumstances."[46] The ILC also acknowledged this priority:

Attaining optimal utilization and benefits . . . implies attaining maximum possible benefits for all watercourse States and achieving the greatest possible satisfaction of all their needs, while minimizing the detriment to, or unmet needs of, each.[47]

In fact, no evidence exists to support the contrary proposition, namely, that waters should be allocated according to the contribution of each state to the basin's waters or according to the length of the river in each state's territory, for example. As Jerome Lipper states unequivocally:

Factors unrelated to the availability and use of waters are irrelevant and should not be considered. For example, the size of a particular state in relation to a co-riparian or the fact that the river flows for a greater distance through one state than another is not in itself a factor to be considered in determining what is an equitable utilization (although it may prove relevant on the issue of "need").[48]

My argument is that application of the "equity of needs" principle is conducive to constructive negotiations. A norm that instructs states to expose not only the natural characteristics of the resource but also their existing and potential needs brings pertinent information to the negotiating table. The discussion over existing and potential needs makes the negotiating riparians aware of the political and other constraints facing their partners and leads them to explore ways to accommodate the interests of all parties. Moreover, comparative needs analysis raises

[46] The Salzburg Resolution, note 15, Article 3.
[47] "ILC Report on the Law of the Non-Navigational Uses of International Watercourses" (1994) 2 *ILC Yearbook* (Part 2), 85, at 97.
[48] Lipper, "Equitable Utilization," note 6, at p. 44.

the potential for domestic support for the negotiated outcome. The allocation and reallocation of transboundary resources such as fresh water and forests involve the vested interests of domestic users. This is not an allocation of previously uncharted continental shelves or maritime resources where no vested interests play a role. Domestic users, especially the strong, agricultural interest groups, will simply resist new allocations that severely curtail their prior uses. Their needs, and their political clout, must be factored into the equation if the outcome is to stand a chance of being ratified and enforced domestically.[49] The primacy of the comparative needs analysis gives content to the vague standard and, thus, an appropriate incentive for states to enter into potentially fruitful negotiations.

Definition of the subject-matter: comprehensiveness versus particularism

Norms of international law can help shape not only the parties' attitudes towards negotiations – whether to negotiate or to litigate – but also, at least to a certain extent, their behavior during the negotiations. If certain claims or positions have no support in international law, a negotiating party is unlikely to present them. However, clear backing in international law lends legitimacy to a claim,[50] and the burden then shifts to the other side to justify its counterclaim to deviate from the norm. While the explanation could be found satisfactory, it could also expose bad faith. Simply put, the default rules of international law have weight – at least some weight – at the negotiating table.[51] This observation suggests that international norms can be instrumental in pushing the negotiating parties towards institution-building rather than towards litigation. The first such default norm is a norm that recommends comprehensive treatment of the ecosystem, as opposed to a norm that endorses particularized treatment of each component of the ecosystem individually.

The story of the evolution of the law on international fresh water and, in particular, the debate in the ILC over the Draft Articles on the

---

[49] On the strong influence of domestic lobbies and the ensuing transnational competition, see the discussion at chapter 3, third part.

[50] See Thomas M. Franck, *The Power and Legitimacy among Nations* (New York, Oxford University Press, 1990), p. 38: "It is the legitimacy of the rules [of international law] which conduces to their being respected."

[51] Franck believes that they carry weight even beyond the negotiations and constrain the discretion of even the most powerful states.

Law on the Non-Navigational Uses of International Watercourses demon-strate the tension between the comprehensive and the particularist ap-proaches. The question at the heart of the debate was how to define the scope of the Draft Articles: Should they apply only to the freshwater bodies – lakes, successive and contiguous rivers – that actually traverse international boundaries? Should they apply also to the entire drainage basin and include all areas within national boundaries that contribute to the flow of the rivers in the basin? And then there was the question of including aquifers and, in particular, the so-called "confined aquifers" – aquifers intersected by international boundaries that do not contribute to the flow of surface waters. The ILC never even considered adopting an "ecosystem approach" that would regard not only the entire basin but also the larger ecosystem as subject to regional regulation.

From a hydrological perspective, compartmentalizing the treatment of fresh water does not make much sense. The hydrological cycle encom-passes clouds, rain, surface water, and underground waters. The entire basin contributes to the flow of both rainwater and effluents that pollute it. Mismanagement of domestic water resources increases a country's reliance on transboundary resources. Thus, a comprehensive approach to the subject-matter of international regulation would be quite obvi-ous to the hydrologist. If the goal is to maximize the benefits from a shared water resource, then the broadest possible definition is called for. A comprehensive definition subjects all water and water-related re-sources whose use could create externality problems to collective regulation. Such resources include lakes, rivers (both successive and con-tiguous), aquifers, and any combination of such resources to which more than one state has access. They include related resources such as forests and aquifer recharge areas that can influence the quantity and quality of the shared water. They also include the atmosphere, which could con-tribute either fumes or additional quantities of rainfall through cloud seeding. In short, from a hydrological point of view, it makes the most sense to adopt a comprehensive, "ecosystem approach" for international regulation.[52]

But a comprehensive approach would be inseparably linked to in-stitutions for transboundary ecosystem management. A comprehensive approach does not lend itself to simple solutions like treaties that appor-tion a river between riparians. Such an approach is clearly detrimental to upper riparians who wish to continue to enjoy unhindered use of

---

[52] See chapter 5, notes 1–4 and accompanying text.

their share of a river or to riparians who have managed in the past to secure a sizable chunk of the river through coercion. It is detrimental to all those who view integrated management of shared resources as an infringement on their sovereignty. It raises obstacles for third parties, such as the World Bank, who seek to fund the projects of only one riparian. Lending institutions will not finance a unilateral project if it is in violation of international law. Because a comprehensive approach narrows the leeway for unilateral action, it also renders more difficult approval for financing projects.

As mentioned, the World Bank has been a key player in the resistance to the adoption of a comprehensive approach. Consistent with its support for the "no harm" rule,[53] the Bank has objected to norms that would render the process of approval of unilateral plans more complex and would require deeper and wider cooperation from all riparians. It therefore has advocated a rather strict definition of the subject-matter for international regulation,[54] preferring adherence to the "traditional" "international river" concept rather than the more comprehensive "international basin" concept suggested by the ILA and IDI. Hence, the Bank has supported governments subscribing to the disengagement philosophy – those that seek to commit their states to as few restrictions as possible under international law. Expressing their preferences at the ILC discussions, these delegates started off with a rather thin definition of the subject-matter for regulation. What was required – according to them – is the regulation of the boundary waters – namely, only those bodies of water that traverse political boundaries – rather than their tributaries.[55] Only later did they concede that the tributaries of the shared rivers also have to be taken into account, and thus the basin approach took root.[56]

In contrast, the basin approach has long been advocated by the IDI and ILA. In the Salzburg Resolution, the IDI mentioned the concept of "hydrographic basin," which is wider than a watercourse.[57] In the 1966

---

[53] See text accompanying notes 35–7.
[54] Krishna, "International Watercourses" note 33, at p. 31, n. 4.
[55] David Goldberg, "World Bank Policy on Projects on International Waterways in the Context of Emerging International Law and the Work of the International Law Commission" in *The Peaceful Management of Transboundary Resources*, note 1, at p. 153.
[56] Ludwik A. Teclaff, "Fiat or Custom: The Checkered Development of International Water Law" (1991) 31 Nat. Res. J. 46, 70–2 (describing the meandering treatment of the issue by the ILC). See also Ludwik A. Teclaff, *Water Law in Historical Perspective* (Buffalo, NY, W. S. Hein, 1985), pp. 424–56.
[57] See Salzburg Resolution, note 15, Article 1.

Helsinki Rules, the ILA formulated a comprehensive definition of the subject-matter for regulation: the "international drainage basin," which encompasses "a geographical area extending over two or more States determined by the watershed limits of the system of waters, including surface and underground waters, flowing into common terminus."[58] Any doubts concerning the inclusion of "confined aquifers" – aquifers whose waters do not "form with surface waters part of a hydrologic system flowing into common terminus" – in this definition were lifted by the 1986 Seoul Rules on International Groundwaters, which clarified that aquifers "intersected by the boundary between two or more States ... constitute an international drainage basin for the purpose of the Helsinki Rules."[59] With the increase in knowledge and experience with respect to the interaction between fresh water and other natural resources, the proposed definition expanded to include the entire ecosystem as the relevant unit for regulation. Thus, as early as 1980, the ILA suggested that

Consistent with Article IV of the Helsinki Rules, States shall ensure that: (a) the development and use of water resources within their jurisdiction do not cause substantial injury to the environment of other States or of areas beyond the limits of national jurisdiction; and (b) the management of other natural resources (other than water) and other environmental elements located within their own boundaries does not cause substantial injury to the water resources of other States.[60]

Moreover, in 1994, the ILA addressed the problem of cross-media pollution, stating, *inter alia*, that "States should co-operate to achieve integrated management of water *and related resources*, including prior assessment of ecological impacts."[61]

The ILC was the forum for airing the controversy between the two visions – integrative and disengagement-oriented – where they clashed

---

[58] Article II of the Helsinki Rules, note 15.

[59] Articles 1 and 2(2) of the ILA's Seoul Rules, in ILA, Report on the Sixty-Second Conference 251 (1987).

[60] Article 1 of the ILA's Articles on the Relationship of International Water Resources with Other Natural Resources and Environmental Elements (Belgrade Conference, 1980, reprinted in ILA, Report of the Fifty-Ninth Conference, Belgrade, 1980, pp. 374–5). In 1996, it was suggested that an integrated treatment of the problems (in relation to the cross-media effects) went beyond the mandate of the ILA's Water Resources Committee (see Harald Hohman, "Articles on Cross-Media Pollution Resulting from the Use of the Waters of an International Drainage Basin" (Report of the ILA's 67th Conference) 412 [1996]).

[61] Article 3 of ILA's report on Cross-Media Pollution, Report of the Sixty-Sixth Conference, Buenos Aires, 1994, at p. 234 (emphasis added).

head-on. The ILC struggled long and hard with the definition of international water resources. This matter had to be deferred until the final stage of the adoption of the Draft Articles on first reading, and then continued to be debated throughout the second reading.[62] The outcome was a definition that only partially encompasses the hydrologically relevant water resources. The ILC only begrudgingly accepted the inclusion of aquifers into the scheme, but excluded "confined aquifers" from its scope. Instead, it issued a "Resolution on Confined Transboundary Groundwater" that, *inter alia*, "commended States to be guided by the principles contained in the [draft articles] where appropriate."[63] Ultimately, the definition adopted by the Watercourses Convention reflects the apprehension with regard to a comprehensive, integrative treatment of the subject-matter, in defining, in Article 2(a), "watercourse" as "a system of surface waters and groundwaters constituting by virtue of their physical relationship a unitary whole and normally flowing into common terminus." This definition does not even mention the basin concept. It excludes confined aquifers for the stated reason that "confined groundwater could be the subject of separate study by the [ILC] with a view to the preparation of draft articles,"[64] implying either that the natural characteristics of such aquifers are yet to be fully understood or that more data are required to assess state practice in this context. The definition is restricted to water flowing to a "common terminus," as though it is crucial that the river flow out into a sea or an ocean, suggesting that rivers that end in the territory of one state are not included in the international regime.[65] Thus, arguably, the Okavango River that rises in the Angolan highlands and flows through Namibia to form the magnificent Okavango Delta in Botswana, as well as the Assi River that flows from Lebanon through Syria into Turkey, are excluded

---

[62] See Stephen S. McCaffrey, Special Rapporteur, *Seventh Report on the Law of the Non-Navigational Uses of International Watercourses*, UN Doc. A/CN.4/436 (1991). This was McCaffrey's final report, devoted entirely to the issue of definition. See also Robert Rosenstock, *Second Report on the Law of the Non-Navigational Uses of International Watercourses*, A/CN.4/462 (1994). For a critical analysis of this debate, see Robert D. Hayton, "Observations on the ILC's Draft Rules: Articles 1–4" (1992) 3 Colo. J. Int'l Env. L. & Poly 31, 34–40.

[63] Resolution on Confined Transboundary Groundwater, Section 1. See the ILC Report, note 47, at p. 135.

[64] See the ILC Report, note 47, at p. 90.

[65] The 1992 Helsinki Convention (note 38) did not adopt the drainage basin approach, but made no distinction between types of aquifers, nor did it refer to the common terminus concept. Article 1(1): " 'Transboundary waters' means any surface or ground waters which mark, cross or are located on boundaries between two or more states..."

because they disappear into the ground in the area of one of the riparians.[66]

The definition is "hydrocentric" in the sense that it plays down the importance of the interaction between water and other media. The Watercourses Convention does not accept the wider notion of shared ecosystems; it does not integrate the entire ecosystem into the considerations of the equitable and reasonable uses; and it does not require states to notify other states on planned measures that will affect the watercourse only indirectly due to effects on the ecosystem.[67] While the Convention does mention, in Article 20, the obligation to "protect and preserve the ecosystems of international watercourses," it does not integrate this obligation into the more minute obligations and processes prescribed in the Convention with respect to measures that affect watercourses. No theoretical explanations of the links that exist between watercourses and the environment are offered, not even in the Preamble to the Convention.

To understand the limitations of this definition, compare it to the following principle set forth in the 1999 Protocol on Water and Health to the 1992 Convention on the Protection and Use of Transboundary Watercourses and International Lakes:

Water resources should, as far as possible, be managed in an integrated manner on the basis of catchment areas, with the aim of linking social and economic development to the protection of the natural ecosystems and of relating water management to regulatory measures concerning other environmental mediums. Such an integrated approach should apply across the whole of a catchment area, whether transboundary or not, including its associated coastal waters, the whole of a groundwater aquifer or the relevant parts of such a catchment area or groundwater aquifer.[68]

This approach is in line with the ICJ decision in *Gabcikovo–Nagymaros*, where it embraced the integrative philosophy and clearly indicated its preference for the comprehensive approach. When referring to Hungary's "right to an equitable and reasonable share of the natural resources of the Danube," the court also examined the damage done to

---

[66] The riparians of the Okavango River do regard that river as shared. In 1994 they formed the Permanent Okavango River Basin Commission (OKACOM). See Stevie C. Monna, "A Framework for International Cooperation for the Management of the Okavango Basin and Delta," Ramsar COP7 DOC. 20.5 (1999) http://www.ramsar.org/cop7_doc_20.5_e.htm.

[67] Articles 11–19 detail duties concerning the exchange of information on planned measures that may affect "the condition of an international watercourse." (Article 11).

[68] Article 5(j), reprinted in http://treaty.un.org/English/notpubl/27-5a-eng.htm.

the "ecology of the riparian area" on the Hungarian side.[69] It stressed "the great significance that it attaches to respect for the environment, not only for States but also for the whole of mankind,"[70] and, therefore, required the two states to take into consideration the wider environmental impacts of their acts. This was a clear endorsement of the comprehensive-integrative philosophy.

Definition of entitlements: shared versus privately owned resources

Another concept that can prove significant for negotiations is the legal definition of the states' control over the transnational resource. As we saw in chapter 2, there is little sense in assigning individual property rights in shares of an international lake or river. Common ownership both reflects reality better and carries greater promise for the evolution of collective action. All common pool transnational natural resources – namely, resources that straddle political boundaries and in which externality problems can occur – should be defined as "international" and as co-owned by the states sharing them.

In this context, too, the drafting process of the Watercourses Convention pitted this integrative concept against the opposite approach, which adhered to the notion of sovereignty – namely, individual ownership over portions of the resource. For a short period of time, between 1980 and 1984, the ILC provisionally accepted the definition suggested by Second Special Rapporteur Stephen Schwebel, who proposed defining these resources as "shared natural resources."[71] But in 1984, following a change in rapporteurs and in the face of rising opposition from certain ILC members, this definition was replaced.[72] The ILC rejected efforts

---

[69] *Gabcikovo–Nagymaros* case, note 3, para. 85. See Paulo Canelas de Castro, "The Judgment in the Case Concerning the Gabcikovo–Nagymaros Project: Positive Signs for the Evolution of International Water Law" (1997) 8 Yb. Int'l Envt'l L. 21, 24.

[70] *Gabcikovo–Nagymaros* case, note 3, para. 53, quoting its previous decision (*Legality of the Threat or Use of Nuclear Weapons*, Advisory Opinion, ICJ Reports 1996, at 241–2, para. 29): "the environment is not an abstraction but represents the living space, the quality of life and the very health of human beings, including generations unborn. The existence of the general obligation of States to ensure that activities within their jurisdiction and control respect the environment of other States or of areas beyond national control is now part of the corpus of international law relating to the environment."

[71] On this concept, see Stephen Schwebel, "Second Report on the Law of Non-Navigational Uses of International Watercourses" reprinted in (1980) 2 *ILC Yearbook* (Part 2), 132–6.

[72] On the change in the ILC position, see McCaffrey, *Second Report*, note 5, at 101–2. See also Patricia Buirette, "Genèse d'un droit fluvial international général" (1991) 95 Rev. Gen. Dr. Int'l Pub. 5, 29–34.

to present the concept of international fresh waters as "shared property." Underlying this hesitation was concern about the implications of such a concept for states' sovereignty and possible undue restrictions on riparians' rights of use. As Max Huber wrote in 1907, "Joint ownership [over transboundary rivers] involves a restriction on the independence of states, and such a restriction may never be presumed, either in regard to state territory itself or with regard to the exercise of territorial sovereignty."[73] The outcome reflects the Convention's philosophy of treating water-related agreements as arm's-length exchanges among sovereign, rather than interdependent, communities.

By the time the Watercourses Convention was adopted, its private-property approach could find support in international legal doctrine. The so-called "community in waters" approach had been rejected as not reflecting positive international law in the *Lac Lanoux* arbitration.[74] There was wide agreement among scholars that while the "community in waters" was the desirable norm, it did not reflect positive law. The law still emphasized sovereign title, as reflected in the 1974 United Nation Charter of Economic Rights and Duties of States (General Assembly Resolution 3281), which subjected unfettered sovereign discretion only to the necessary limitations imposed by the principle of good neighborliness and the doctrine of abuse of rights (*sic utere tuo ut alienum non laedas*).[75]

But Professor – now Judge – Schwebel and his colleagues on the ICJ bench turned out to have the final say. The ICJ took advantage of the *Gabcikovo–Nagymaros* dispute as an opportunity to infuse the integrative philosophy into the Watercourses Convention, thereby overcoming the stasis of the doctrine. In the decision, the ICJ referred to the Convention as endorsing the concept of "community of interest in a navigable river,"

---

[73] See Max Huber, "Ein Beitrag zur Lehre von der Gebietshoheit an Grenzflüssen" in (1907) *Zeitschrift für Völkerrecht und Bundesstaatsrecht*, translated and cited in Berber, *Rivers*, note 6, at p. 25.

[74] This is the so-called "community in waters" approach: *Lac Lanoux* Arbitration, (1957) 24 Int. L. R. 101; Lipper, "Equitable Utilization," note 6, at pp. 38–40; Berber, *Rivers*, note 6, at pp. 22–5; Buirette, "Genèse," note 72, at p. 34.

[75] See Article 3 of the Resolution: "In the exploitation of natural resources shared by two or more countries, each State must co-operate on the basis of a system of information and prior consultations in order to achieve optimum use of such resources without causing damage to the legitimate interests of others." This is the so-called "limited territorial sovereignty" approach. See, e.g., Lipper, "Equitable Utilization" note 6, at pp. 23–38; Berber, *Rivers*, note 6, at pp. 22–41; Marc Wolfrom, *L'utilisation à des fins autres que la navigation des eaux des fleuves, lacs et canaux internationaux* (Paris, A. Pedone, 1964), pp. 63–7. This is also the approach taken by the ILC Special Rapporteur McCaffrey, *Second Report*, note 5, at pp. 130–1, who referred to this doctrine as "the bedrock upon which the doctrine of equitable utilization is founded."

which is "the basis of a common legal right," and applying it to trans-boundary water resources.[76] This concept was first enunciated by the Permanent Court of International Justice in 1929 in the *River Oder* case, with respect to freedom of navigation in rivers.[77] Moreover, the ICJ added that the Watercourses Convention accepted "the concept of common utilization of shared water resources."[78] This was a rather bold and dramatic move, particularly in light of the disapproving position explicitly expressed in the Watercourses Convention in this respect.

The significance of the victory garnered by the integrative camp on this matter of characterization of title should not be underestimated. Such characterization does not, in itself, change riparians' rights and obligations. But it does enhance the legitimacy of the integrative vision. It may also have a bearing on the direction of the development of the default rules on interstate cooperation. This characterization could lead to a shift from a Lotus-based approach, where limitations on state sovereignty must be proven, to just the opposite approach, where the burden of proof is borne by the side that seeks to avoid responsibility and cooperation.

Identifying the stakes: the triumph of the human over the statist dimension

Herbert Arthur Smith's pioneering book, *The Economic Uses of International Rivers*, published in 1931 and the first treatise on transboundary resources,[79] was written during a period when the challenges raised by this subject-matter were mainly economic in nature. As Smith wrote in the introduction, the aim of the book was to address the legal aspects of diversions and other "artificial interferences with the natural course of streams . . . directed to various economic ends, such as navigation, irrigation, and the development of hydro-electric power, [which] in many cases . . . have given rise to a serious conflict of *state interests*."[80] Indeed,

---

[76] *Gabcikovo–Nagymaros* case, note 3, para. 85.

[77] *Territorial Jurisdiction of the International Commission of the River Oder*, Judgment No. 16, 1929, PCIJ, Series A, No. 23, at 27.

[78] *Gabcikovo–Nagymaros* case, note 3, para 147: "Re-establishment of the joint regime will also reflect in an optimal way the concept of common utilization of shared water resources for the achievement of the several objectives mentioned in the [1977] Treaty, in concordance with Article 5, paragraph 2 of the [Watercourses Convention]."

[79] Herbert Arthur Smith, *The Economic Uses of International Rivers* (London, P. S. King & Son, Ltd., 1931).

[80] Smith, *Economic Uses*, at p. v (emphasis added).

until relatively recently, the concept of use of water resources did not, in most cases, encompass human consumption.[81]

Smith's characterization of water conflicts as economic conflicts among states was the result of two factors. First, when Smith wrote his book, international law focused solely on national interests and did not allow individual or communal rights into its purview. Second, at that time, the international and interstate disputes that reached legal or diplomatic settlements reflected purely economic interests, rather than immediate threats to human survival. For example, competition between Belgium and The Netherlands for the development of the ports of Antwerp and Rotterdam lay at the heart of conflicting unilateral plans for constructing canals to divert waters out of the shared Meuse River.[82] Illinois' plans to divert water from Lake Michigan into the Mississippi, aimed at disposing sewage from the Chicago area (and, later, at harnessing the flow for hydroelectric power), was objected to as adversely affecting Canadian economic interests in navigability on the St. Laurence River from Montreal to the Atlantic Ocean. The use of river flows for generating power was at the center of the dispute between the cantons of Aargau and Zurich in Switzerland (between 1871 and 1878); while the diversion of the Danube by Baden was contested in the German Courts by downstream Wuerttemberg due to the loss of economic opportunities.[83]

But as conditions of scarcity developed in many basins, competition for water became a matter not only of economics for states. In a growing number of basins, competition for water has become more and more about human subsistence and the provision of basic human needs and rights; it relates not only to individuals, but also affects the preservation of communities and cultures.

Because water affects individuals and communities in so many different ways, developments in different spheres of law have provided for indirect, albeit important, impact on the law of watercourses. Emerging principles of international environmental law and of human rights law provide basic norms to which all arrangements must conform. The law on armed conflicts imposes restrictions on the use of access to water as a weapon. The laws on treaties and on state succession govern states'

---

[81] Berber, *Rivers*, note 6, at p. 5.
[82] This competition began with the separation of Belgium from the Dutch Kingdom in 1833 and was resolved only in 1933. The use of the Rio Grande River for irrigation on both sides of the border lay at the heart of the dispute between Mexico and the United States from 1895 to 1906.
[83] See Smith, *Economic Uses*, note 79.

undertakings under treaties, and principles of state responsibility determine the responsibility of states for acts of individuals related to the use
of water within their territories. Finally, developments in international
trade law have bearing on states' attitudes towards "food security" and
similar constraints that inform their decisions with respect to water
allocations.[84] Any assessment of the current law on freshwater resources
must analyze the confluence of these disparate areas of the law.

In this sense, the Watercourses Convention covers only some of the
aspects of the international body of laws relating to international
watercourses. It continues to view the matter as an essentially interstate conflict. Individual and group rights, the representation of both in
the decision-making and negotiating processes as well as in the actual
outcome of those processes, are not addressed by the Convention.

As to be expected, the ICJ takes a different approach. The court acknowledged the human dimension of this issue in its Advisory Opinion on the *Legality of the Threat or Use of Nuclear Weapons*, stating that
"the environment is not an abstraction but represents the living space,
the quality of life and the very health of human beings, including
generations unborn."[85] This statement was quoted by the court in
its *Gabcikovo–Nagymaros* decision,[86] while Vice President Weeramantary
added in his separate opinion that "the people of both Hungary and
Slovakia are ... entitled to the preservation of their human right to the
protection of their environment" and that "the protection of the environment is ... a vital part of contemporary human rights doctrine, for
it is a sine qua non for numerous human rights such as the right to
health and the right to life itself."[87]

Identifying the goals of negotiations: dispute settlement versus establishment of institutions

In 1931, Herbert Arthur Smith's book identified correctly the potential
and limitations of international law in promoting cooperation and resolving disputes between riparian states. Smith warned against attempts
to find "a real solution to these water problems by reference to purely
legal tribunals which can only decide them by applying abstract legal
rules."[88] He stated that "there is really no general rule of law which
can be applied indiscriminately to all disputes that may arise. Each

---

[84] On the concern with food security see chapter 5, notes 90–1 and accompanying text.
[85] ICJ Reports 1996, para. 29.   [86] *Gabcikovo–Nagymaros* case, note 3, paras. 53, 118.
[87] *Ibid.* (Judge Weeramantary's opinion).   [88] See note 79, at p. 57.

river system requires to be separately considered in light of its own history and physical conditions."[89] Rather than the formulation of dispute-settlement-oriented legal rules, Smith identified the primary challenge as follows: "The main problem presented by the economic development of international rivers is not so much the formulation of legal rules as the constitution of authorities which shall settle disputed questions."[90] The importance of shared regimes for the management of transboundary resources was recognized even before Smith's book saw the light. Already in 1911, the sages at the Institut de droit international (IDI) concluded in their Madrid Resolution that

II. When a stream traverses successively the territories of two or more states:
...
g. It is recommended that the interested states appoint permanent joint commissions, which shall render decisions, or at least give their opinions, when new establishments or modifications of existing establishments may cause significant consequences for the party of the watercourse situated on the territory of the other state.[91]

Smith was correct in emphasizing not only that joint commissions are beneficial, but that it would be futile to concentrate efforts on formulating specific substantive rules for the allocation of a shared resource. Needless to say, the knowledge that has accrued since Smith's book, analyzed in the previous chapters, has only strengthened his observation.

Emphasis on joint management seemed to prevail in subsequent legal literature. Fifty years after its first Resolution, the IDI reiterated the importance of joint management in its Salzburg Resolution, which recommended "recourse to technical experts . . . in order to arrive at solutions assuring the greatest advantage to all concerned."[92] Joint management was lauded by the ILA in 1986;[93] and a number of multilateral conventions have actually prescribed a duty of joint management.[94]

---

[89] Smith, *Economic Uses*, at p. 87.   [90] *Ibid.*, at p. 120.

[91] In the original French: "Il est recommandé d'instituer des commissions communes et permanentes des états intéressés qui prendront des décisions, ou tout au moins donneront leur avis, lorsqu'il se fera des nouveaux établissements ou des modifications aux établissements existants pour la partie du cours d'eau située sur le territoire de l'autre état" (1911) 24 Ann. Inst. Dr. Int'l 365.

[92] This is the IDI's Preamble to its 1961 Salzburg Resolution, note 15.

[93] See the Seoul Rules, note 59, Article 4: "Basin States should consider the integrated management, including conjunctive use with surface waters, of their international groundwater at the request of any one of them."

[94] The Helsinki Convention, note 38, Article 9 (2) (the agreements between the riparian parties "shall provide for the establishment of joint bodies," which are defined as

State practice, observed mainly through bilateral and regional treaties, has indicated a growing awareness of the benefits of institutional cooperation.[95]

During the lengthy process towards drafting the Watercourses Convention, efforts were made to recognize the importance of institutional structures for shared management of transboundary resources. Special Rapporteur Stephen McCaffrey put forward one elaborate suggestion.[96] Commentators developed rather complex procedures for shared management[97] and unanimously criticized the insufficient text of Draft Article 24(1) with regard to "management."[98]

However, despite this criticism, Draft Article 24(1) became the final Article 24(1) of the Watercourses Convention, without any change in formulation. Under this article, "watercourse States shall, at the request of any of them, *enter into consultations* concerning the management of an international watercourse, which *may* include the establishment of a joint management mechanism."[99] At the same time, however, a new provision was added to Article 8, whose heading is "General obligation

"any bilateral or multilateral commission or other appropriate institutional arrangements for cooperation" [Article 1]); the ECE Charter on Groundwater Management (ECE Annual Report (1989–90), ECOSOCOR 1989, Supp. No. 15), Article 25(1): "Concerted endeavours to strengthen international co-operation for harmonious development, equitable use and joint conservation of ground-water resources, located beneath national boundaries, should be intensified . . . In order to implement such co-operation, joint commissions or other intergovernmental bodies should be established. . . . ."

[95] For surveys of such institutions, see, for example, Richard D. Kearney, "International Watercourses" in René-Jean Dupuy (ed.), *A Handbook on International Organizations* (Dordrecht, M. Nijhoff, 1988), pp. 509–58; Brice M. Clagett, "Survey of Agreements Providing for Third-Party Resolution of International Water Disputes" (1961) 55 AJIL 645, 655, 660–1.

[96] Stephen S. McCaffrey, *Sixth Report on the Law of the Non-Navigational Uses of International Watercourses*, Addendum, UN Doc. A/CN.4/427/Add.1 (1990). For an analysis of this suggestion and the ILC Draft on this matter, see Sergei V. Vinogradov, "Observations on the ILC's Draft Rules: 'Management and Domestic Remedies'" (1992) 3 Colo. J. Int'l Envt'l L. & Pol'y 235, 235–41.

[97] See the Bellagio Draft Treaty, note 17.

[98] For criticism of one of the Draft's "weakest provisions," see Constance D. Hunt, "Implementation: Joint Institutional Management and Remedies in Domestic Tribunals (Articles 26–28 and 30–32)" (1992) 3 Colo. J. Int'l Envt'l L. & Pol'y 281, 284–7.

[99] Article 24(1), note 2 (emphasis added). Article 24(2) defines "management," in particular, as "(a) Planning the sustainable development of an international watercourse and providing for the implementation of any plans adopted; and (b) Otherwise promoting the rational and optimal utilization, protection and control of the watercourse."

to cooperate" and whose section (1) stipulates a duty to cooperate "to attain optimal utilization and adequate protection" of international watercourses. The new section (2) reads: "In determining the manner of such cooperation, watercourse States may consider the establishment of joint mechanisms or commissions, as deemed necessary by them, to facilitate cooperation on relevant measures and procedures in the light of experience gained through cooperation in existing joint mechanisms and commissions in various regions."[100] This is a marked improvement over the Draft Articles and Article 24, because it links the idea of joint mechanisms to both the obligation to cooperate and the goal of optimal utilization and adequate protection. Article 24 does not make any such connection. Its hesitant and elliptical endorsement of such joint mechanisms (referring to the experience gained in existing commissions) could have been firmer and more direct. As it stands, Article 24 presents joint mechanisms as merely an afterthought to the general obligation to cooperate, only one of the options available to riparians for fulfilling their obligation to cooperate. Compared with the detailed treatment of procedural rules on the exchange of data and information on a regular basis (Article 9) and of the obligations with respect to planned measures (Part III, Articles 11–19)[101] and with the recourse to impartial fact-finding or conciliation in dispute settlement (Article 33),[102] Article 8(2) emerges as a rather weak endorsement of what should have been one of the pillars of the Convention.

This reluctance to endorse joint management is all the more puzzling in view of the non-compulsory nature of the Watercourses Convention, which is only a framework convention and, hence, not immediately binding on negotiating riparians.[103] Moreover, the Convention does make the

---

[100] Article 8(2), note 2.

[101] For criticism of the 1991 ILC Draft for failing to provide satisfactory procedural rules for the exchange of information and notification of planned measures, see Bourne, "The Right to Utilize", note 26, at pp. 68–72; Alberto Szekely, " 'General Principles' and 'Planned Measures' Provisions in International Law Commission's Draft Articles on the Non-Navigational Uses of International Watercourses: A Mexican Point of View" (1992) 3 Colo. J. Int'l Envt'l & Pol'y 93, 99–100; Gunther Handl, "The International Law Commission's Draft Articles on the Law of International Watercourses (General Principles and Planned Measures): Progressive or Retrogressive Development of International Law?" (1992) 3 Colo. J. Int'l Envt'l & Pol'y 123, 124–9.

[102] At the Sixth Committee, many commentators suggested the inclusion of binding procedures for arbitration and judicial settlement; many others, however, rejected that idea. *Summary Records of the Meetings of the Sixth Committee*, A/C.6/49/SR.17-26 (1994).

[103] The proposed convention has a dual character: many of its provisions are nonbinding, yet at the same time, its provisions aim to reflect existing law. The text

bold move of imposing an obligation on riparians – the obligation to re-sort to a fact-finding commission in the event of conflict – but this is an obligation that sets the parties down the wrong track, towards litigation rather than negotiation. With its uninspiring treatment of management, on the one hand, and its detailed provisions for dispute settlement, on the other, as well as its provision on equal access to national courts (Article 32), the Convention sends out the wrong signal to politicians and lawyers: that litigation as a means of securing optimal and equi-table use of water resources is preferable to negotiations leading to joint management institutions.[104]

In stark contrast, the ICJ declared unequivocally its preference for strong regional joint management institutions, institutions that "reflect in an optimal way the concept of common utilization of shared water resources."[105] The *Gabcikovo–Nagymaros* decision compelled the par-ties to resume negotiations in earnest, rather than wait for a judicial fiat:

It is not for the court to determine what shall be the final result of these negoti-ations to be conducted by the Parties. It is for the Parties themselves to find an agreed solution that takes account of the objectives of the Treaty, which must be pursued in a joint and integrated way, as well as the norms of international en-vironmental law and the principles of the law of international watercourses.[106]

In the same decision, the ICJ concluded by emphasizing joint manage-ment institutions as the preferred modality for managing transboundary resources:

Re-establishment of the joint regime will also reflect in an optimal way the concept of common utilization of shared water resources for the achievement of the several objectives mentioned in the Treaty, in concordance with Article 5, paragraph 2, of the [Watercourses Convention].[107]

---

is therefore unclear on the rights and obligations of riparians absent an agreement. On this issue, see the *different* views of Mr. Caflisch (Switzerland), Mr. Ridruejo, (Spain), Ms. Skrk (Slovenia), and Ethiopia at the Sixth Committee, *Summary Records of the Meetings of the Sixth Committee*, A/C.6/49/SR.22, at 15; SR.23, at 10; SR.26, at 8; SR.28, at 2 (respectively).

[104] Indeed, while the Draft Article on dispute settlement drew much attention at the Sixth Committee, the provision on management did not provoke any reaction. There were only two comments on the issue of joint management, in the context of Draft Article 5. Both comments supported the idea. See Summary Records, *supra* note 103, SR.24, at 8, 12–13 (Mr. Elarabi, Egypt, and Mr. Momtaz, Iran, respectively).

[105] *Gabcikovo–Nagymaros* case, note 3, para. 147.

[106] *Ibid.*, para. 141.   [107] *Ibid.*, para. 147.

The court also pointed out that this conclusion follows from the law on state responsibility:

> In this case, the consequences of the wrongful acts of both Parties will be wiped out "as far as possible" if they resume their co-operation in the utilization of the shared water resources of the Danube, and if the multi-purpose pro-gramme, in the form of a co-ordinated single unit, for the use, development and protection of the watercourse is implemented in an equitable and reasonable manner.[108]

Institution-enhancing norms

The second sphere in which international law can create the proper incentives for states to opt for institutional arrangements is that of default norms that conform with the principles for sound management of the transnational resources detailed in chapter 6. Although riparians can deviate from these norms by mutual consent, setting the default rules to conform with the theory on collective action transforms the legal environment from a Lotus-based system to an integrative one. Such default settings provide guidance for negotiators who aim for cooperation and increase the transaction costs for those who seek other goals.

This section examines the status of international norms relevant to the functioning of transnational ecosystem institutions. It picks up from the discussion in chapter 6 on the constraints and tasks of such institutions as well as the possible mechanisms for performing those tasks.

Supremacy of institutional policies

As discussed throughout the previous chapters, long-term cooperation, which is the key to sustainable and optimal resource management, may face domestic challenges from groups pressuring their governments to renege on their treaty obligations. This requires that a high threshold be raised for a unilateral right to exit from joint institutions, which is both appropriate and in the long-term interest of the participating states. All things being equal, the stricter the rules precluding unilateral exit from treaty obligations, the stronger state parties' commitment to long-term cooperation, and the less the uncertainty as to possible future breaches. Two other principles contribute towards the goal of fortifying commitment to long-term cooperation. The one principle restricts the

---

[108] *Ibid.*, para. 150.

effects of state succession on the transnational institution; the other restricts the weight assigned to the property rights of individuals in specific shares of the resource and makes these rights subject to the authority of the institution.

The analysis of the ICJ decision in the *Gabcikovo–Nagymaros* case indicates that the court was motivated by the same policy of strengthening international commitments at the expense of domestic interests. The decision clearly raised the hurdle for unilateral exit, with the court ruling that the project, despite its huge dimensions and irreversible consequences and despite the fact that its architects had been the long-gone Hungarian and Czech communist regimes, must continue and could only be changed with the consent of both parties.[109] The court apparently foresaw its instruction as arming governments in future disputes against domestic opposition to international agreements concerning transboundary resources, since any alternative government would be likewise bound by the obligations thereunder.

For these same reasons, transnational institutions must be immune to the effects of state succession. It is generally accepted that treaties concerning the determination of water rights or navigation on rivers are "territorial" and thus continue to be in force with succession.[110] Under the 1978 Vienna Convention on Succession of States in respect of Treaties,

A succession of States does not as such affect: (a) obligations relating to the use of any territory, or to restrictions upon its use, established by a treaty for the benefit of a group of States or of all States and considered as attaching to that territory; (b) rights established by a treaty for the benefit of a group of States or of all States and relating to the use of any territory, or to restrictions upon its use, and considered as attaching to that territory.[111]

In the *Gabcikovo–Nagymaros* decision, the ICJ found this Article to reflect customary international law and therefore decided that the contents of the Czechoslovak–Hungarian 1977 treaty "must be regarded as establishing a territorial regime within the meaning of Article 12 . . . It created

---

[109] For a description of the case, see chapter 6, notes 9–21 and accompanying text.

[110] Daniel P. O'Connell, *State Succession in Municipal and International Law* (Cambridge, Cambridge University Press, 1967), vol. II, at p. 273; Jose C. F. Rozas, "La succession d'Etats en matière de conventions fluviales" in Ralph Zacklin and Lucius Caflisch (eds.), *The Legal Regime of International Rivers and Lakes* (The Hague, M. Nijhoff, 1981), pp. 127, 158–60; Yimer Fisseha, "State Succession and the Legal Status of International Rivers" in Zacklin and Caflisch, *Legal Regime* at 177. See also "Succession of States in respect of Bilateral Treaties: Study Prepared by the Secretariat" (1970) 2 *ILC Yearbook* 102, 111–12.

[111] Article 12(2).

rights and obligations 'attaching to' the parts of the Danube to which it relates; thus the Treaty itself cannot be affected by a succession of States."[112] In coming to this conclusion, the court clearly indicated that treaties providing for the construction and joint operation of a large, integrated, and indivisible complex of structures and installations on specific parts of the territories of the riparian parties would continue to operate despite succession.[113]

The competence of the transboundary ecosystem management institution to interfere in the domestic allocation of entitlements is another aspect of the supremacy of institutional policies. This may involve domestic assignment of property rights to individuals, rights that may be constitutionally protected and, hence, difficult to alienate. From the institution's perspective, there should be no legal impediment to shifting entitlements among individual users and administrative costs should be low. Considerations of private ownership should not hinder decision-makers seeking to modify allocations in achieving efficient and equitable outcomes.

Two principles support this policy. The first relates to the protection of private entitlements as human rights. The right to property is not recognized as a human right in international legal instruments, and it is generally viewed as subject to takings for public purposes.[114] Hence, under international law there is no legal obstacle to modifying existing entitlements of individuals. However, despite this absence of any impediment to modification, there may, nevertheless, be good reasons for granting precedence to existing uses in a given shared resource. In this context, the second principle, which relates to the status of existing uses vis-à-vis future uses of shared fresh water, gains relevance. There are two positions on this issue. The 1966 Helsinki Rules accord conditional priority to existing uses. Article 8(1) of the Helsinki Rules states: "An existing reasonable use may continue in operation unless the factors justifying its continuance are outweighed by other factors leading to the conclusion that it be modified or terminated so as to accommodate a competing incompatible use."[115]

This principle was also recognized by the US Supreme Court,[116] which placed a heavy burden on the party challenging current allocations to

---

[112] See note 3, para. 123.   [113] *Ibid.*

[114] The right to private property is not included in the 1966 Covenant on Civil and Political Rights. As to the right to property under the 1950 European Convention, see Jochen Abr. Frowein, "The Protection of Property" in R. St. J. Macdonald *et al.* (eds.), *The European System for the Protection of Human Rights* (Dordrecht, M. Nijhoff, 1993), p. 515.

[115] In the same vein, see the Salzburg Resolution of the IDI, *supra*, note 15, Articles 3, 4.

[116] *Colorado* v. *New Mexico*, 459 US 176, 103 S. Ct. 539, 74 L.Ed.2d 348 (1982).

prove the desirability of modification. In its view, "the equities supporting the protection of existing economies will usually be compelling."[117]
In clear and deliberate contrast, the Watercourses Convention[118] assigns
existing uses the same weight as potential ones.[119] While there are good
policy reasons for assigning conditional priority to existing uses,[120] it
is clear that such priority is a far cry from total immunity against
modification.

Ensuring the flexibility of institutions

It is crucial that transboundary resources-related treaties enjoy relative
flexibility in order to face the challenge of accommodating changing
conditions, in terms of both supply of and demands on the shared
resource.[121] Thus, ecosystem management agreements should be subject
to standards that are more flexible than the rather strict rules that apply
to discrete agreements. Ecosystem management agreements should be
interpreted and implemented with the necessary flexibility, even with respect to existing agreements that specify rigid allocation of shares. The
flexible standard of "equitable and reasonable use" that was found to
be conducive in the negotiations phase is also instrumental during the
lifetime of the institution. The standard should be understood as permitting reallocations on a periodic basis, without requiring that the party
seeking modification of the allocations be bound to the strict doctrine
of *rebus sic stantibus*.[122] Application of the "equitable and reasonable use"

[117] *Ibid.*, 459 US at 187. To justify the detriment to existing uses, a state would have to
"demonstrate by clear and convincing evidence that the benefits of the [change]
substantially outweigh the harm that might result" (*ibid.*). See also Richard A. Simms,
"Equitable Apportionment – Priorities and New Uses" (1989) 29 Nat. Res. J. 549; A.
Dan Tarlock, "The Law of Equitable Apportionment Revisited, Updated, and Restated"
(1985) 56 U. Col. L. Rev. 381.

[118] See the commentary to the ILC draft Article 7, *ILC Yearbook – 1987*, vol. II, part 2, at 36
(1989).

[119] Article 6(1)(e) couples "existing and potential uses" and further on stipulates that
neither of the factors enjoys precedence (Article 6(3)).

[120] See chapter 6, notes 34, 36 and accompanying text.

[121] See chapter 5, notes 80–1, chapter 6, notes 34–8 and accompanying text.

[122] Under Article 62(1) of the 1969 Vienna Convention on the Law of Treaties, a party
may unilaterally withdraw from its treaty obligations under very strict conditions.
The party must show, *inter alia*, that "a fundamental change of circumstances has
occurred with regard to those existing at the time of the conclusion of the treaty,
and which was not foreseen by the parties." Usually, changes in demand or supply of
transboundary resources would be incremental, quite foreseen by the parties to the
agreement over the initial allocation. See also chapter 6, note 8 and accompanying
text.

standards means that all allocations are subject to future adjustment: whenever an allocation becomes inequitable or unreasonable, the standard mandates reallocations to adapt to the new circumstances. This standard does not assign property rights in water shares, but, rather, rights subject to reevaluation based on an assessment of concrete criteria. In the renegotiation of allocations, existing beneficial uses are granted only qualified priority. This conditional status helps to ensure flexibility and mutuality among the riparians in their future interactions and, hence, creates incentives for the parties to undertake long-term commitments and cooperate.

In this respect, the Watercourses Convention is a disappointment. First, it does not recognize the need for flexibility and sensitivity to changes external to agreements already in force. Article 3 stipulates that the Convention will not affect existing agreements and adds that notwithstanding this principle, parties to existing agreements "may, where necessary, consider harmonizing such agreements with the basic principles" of the Convention.[123] Second, the Convention is silent on the subject of subsequent modification of water-related agreements, implying the applicability of the general doctrine of *rebus sic stantibus*.

In this context, too, it was the ICJ decision in *Gabcikovo–Nagymaros* that set legal development on its proper course. By reading into the 1977 treaty a flexibility that opens the way to renegotiating the treaty's basic provisions in light of new developments in international law, the court introduced a new understanding of environmental impacts and new circumstances. The court upheld the 1977 treaty only after reading into it flexibility and mutuality and after emphasizing the parties' duty to achieve the object and purpose of the developing treaty relationship (an object and purpose that the court in fact postulated in light of developing international law).[124] Whereas a discrete agreement would have assigned finite rights and obligations to the parties, the court refused to view a watercourse-related agreement as producing such finite consequences. Instead, it stressed the continuous and evolving character of such agreements, which must remain sensitive to new scientific

---

[123] Article 3(1) and 3(2), note 2. Note also that one of the "statements of understandings pertaining to certain articles of the convention," regarding Article 3, provides that the Convention "will serve as a guideline for future watercourse agreements and, once such agreements are concluded, it will not alter the rights and obligations provided therein, unless such agreements provide otherwise."

[124] See note 3, paras. 132–47. Judge Bedjaoui criticized this evolutionary interpretation; see his separate opinion at http://www.icj-cij.org/idocket/ihs/ihsjudgement/ihsjudframe1.htm.

insights, stating, "The awareness of the vulnerability of the environment and the recognition that environmental risks have to be assessed
on a continuous basis have become much stronger in the years since
the [1977] Treaty's conclusion."[125] New developments in international
law relating to the work of transnational institutions also must be considered. Thus the court found the parties obliged to take into account
"newly developed norms of environmental law [that] are relevant for
the implementation of [the 1977] Treaty."[126] In one of the clearest ex-
pressions of the cooperative philosophy, the court declared,

Throughout the ages, mankind has, for economic and other reasons, constantly
interfered with nature. In the past, this was often done without consideration
of the effects upon the environment. Owing to new scientific insights and to a
growing awareness of the risks for mankind – for present and future generations – of pursuit of such interventions at an unconsidered and unabated pace,
new norms and standards have been developed, set forth in a great number
of instruments during the last two decades. Such new norms have to be taken
into consideration, and such new standards given proper weight, not only when
States contemplate new activities but also when continuing with activities begun
in the past. This need to reconcile economic development with protection of the
environment is aptly expressed in the concept of sustainable development.[127]

Norms concerning the operation of the institutions

In chapter 6, a number of elements crucial for the proper functioning
of transnational ecosystem institutions were identified. This section will
examine to what extent prevailing international law provides support
for these elements.

Data collection and dissemination

The recognition of the duty of governments to provide information to
their counterparts and to the joint institution could, technically, be
based on the duty to cooperate in the utilization of shared resources
or on the more general duty to negotiate resource-related agreements
in good faith. Because resource-sharing agreements focus on questions
of management and allocation of the resource on the basis of the respective supplies and demands of the parties, withholding substantial
and accurate information on these two points runs contrary to the duty

[125] See note 3, para. 112.   [126] *Ibid.*   [127] See note 3, para. 140.

to cooperate and constitutes bad faith. Thus, it is widely accepted that
"watercourse States shall exchange information."[128]

Parallel to the collection of information in the framework of the institution lies the function of dissemination of the same information to the
general public. A duty to do so can be derived from the principle of freedom of information, a widely accepted principle in many democracies
and deeply rooted in international human rights law.[129] Such a duty
has support in international instruments that address transboundary
ecosystems. The 1992 Helsinki Convention on the Protection and Use of
Transboundary Watercourses and International Lakes, for example, requires riparians to "ensure that information on the conditions of the
transboundary waters, measures taken or planned to be taken to prevent, control or reduce transboundary impact, and the effectiveness of
those measures, is made available to the public."[130] As the 1999 Protocol
on Water and Health to the 1992 Convention explains: "Access to information and public participation in decision-making concerning water
and health are needed, *inter alia*, in order to enhance the quality of the
implementation of the decisions, to build public awareness of issues, to
give the public the opportunity to express its concerns and to enable
public authorities to take due account of such concerns."[131]

Public participation

The benefits of public participation have been recognized in recent international instruments. The 1992 Rio Declaration notes that "environmental issues are best handled with the participation of all concerned
citizens, *at the relevant level*."[132] Chapter 18 of Agenda 21 calls for active

---

[128] Article 11 of the Watercourses Convention, note 2.

[129] On transparency as a crucial principle in the work of the European institutions, see
Paul P. Craig and Grainne de Burca, *EC Law: Texts, Cases and Materials* (Oxford,
Clarendon Press, 1995), pp. 368–71; Veerle Deckmyn and Ian Thompson, *Openness and
Transparency in the European Union* (Maastricht, The Netherlands, European Institute of
Public Administration, 1998).

[130] Article 16. See also Agenda 21, chapter 18 (on freshwater resources), Principle
18.12(p) (concerning the dissemination of information as one of the means of
improving integrated water management), reprinted in Nicholas A. Robinson (ed.),
Agenda 21 & the UNCED Proceedings (6 vols., New York, Oceana Publications, 1992), vol.
IX, pp. 357, 366.

[131] Article 5 (I), *supra* note 68. See also the Aarhus Convention on Access to Information,
Public Participation in Decision-Making and Access to Justice in Environmental
Matters 1998 (1999) 38 ILM 517.

[132] Declaration of the UN Conference on Environment and Development, Rio de Janeiro,
3–14 June 1992, Doc A/CONF 151/5/Rev.1 (1992) Principle 10 (emphasis added).

public participation in shared freshwater management. This includes not only the provision of a right of hearing to oppose plans that could be detrimental to certain individuals or groups, but, more generally, the obligation of states to aim for "an approach of full public participation, including that of women, youth, indigenous people and local communities in water management policy-making and decision making,"[133] suggesting the "development of public participatory techniques and their implementation in decision-making."[134] A number of conventions grant opportunities for public participation through NGOs as observers.[135] The NAFTA agreement goes even further, providing standing for NGOs to complain about a state's failure to comply with its obligations.[136] The 1997 Helsinki Declaration of the state parties to the 1992 Helsinki Convention on the Protection and Use of Transboundary Watercourses and International Lakes states that "broad public participation is essential for implementing and developing further the convention."[137] The Declaration mentions governments, public and private sector organizations, joint bodies, NGOs, the scientific community, and all those involved in water management and environmental protection as potential participants in the process.

The 1997 Watercourses Convention fails to embody this idea of public participation. It maintains a strict division between the international and domestic levels by providing only for state-to-state notification and consultation. There are, however, scholars who derive such participatory rights from basic notions of civil and political rights[138] and of general environmental law, particularly if the right to a clean and healthy environment and fresh water is considered within the scope of the individual's human rights.[139]

[133] *Agenda 21*, note 130, Principle 18.9(c).

[134] *Agenda 21*, note 130, Principle 18.12(n). See also Ellen Hey, "Sustainable Use," note 1.

[135] Treaties that provide standing to NGOs as observers include the 1987 Montreal Protocol on Substances that Deplete the Ozone Layer (Article 11), the 1992 Framework Convention on Climate Change (Article 7) and the 1999 Bern Convention on the protection of the Rhine (Article 14).

[136] NAFTA's North American Agreement on Environmental Cooperation (including standing to complain against a state for failing to enforce domestically); see Kal Raustiala, "International 'Enforcement of Enforcement' under the North American Agreement on Environmental Cooperation" (1996) 36 Va. J. Int'l L. 721.

[137] Report of the First Meeting, ECE/MP.WAT/2, at 17.

[138] See Alan Boyle, "The Role of International Human Rights Law in the Protection of the Environment" in Alan E. Boyle and Michael R. Anderson (eds.), *Human Rights Approaches to Environmental Protection* (Oxford, Clarendon Press, 1996), pp. 43, 59.

[139] See Boyle, *International Human Rights Law* at 59–63; James Cameron and Ruth Mackenzie, "Access to Environmental Justice and Procedural Rights in International

Representation of minorities

Of particular importance is the participation of minority groups in decisions that have the potential to affect their interests. These groups tend to be underrepresented in the national decision-making process, and hence, their perspectives vis-à-vis their needs and equitable and reasonable use of a given resource may be lost in the national arena. It is therefore crucial to give them voice in the transnational arena. This has been recognized by the UN Human Rights Committee, which set forth the duty of member states to take "measures to ensure the effective participation of members of minority communities in decisions which affect them."[140] Minority groups' "right to participate effectively" in public life and in matters concerning them is recognized in the 1995 Council of Europe's Framework Convention for the Protection of National Minorities,[141] as well as in the Declaration on the Rights of Persons Belonging to National or Ethnic, Religious and Linguistic Minorities.[142] State practice in the recently established democracies in Central and Eastern Europe reflects a similar endeavor to ensure minority representation in democratic processes, for example, in parliament. These efforts include constitutional guarantees[143] as well as constitutional court decisions protecting minority groups' right to participation.[144]

Institutions" in *Human Rights Approaches*, note 138, at 129 (esp. 134–5); Sionaidh Douglas-Scott, "Environmental Rights in the European Union – Participatory Democracy or Democratic Deficit?" *in Human Rights Approaches*, note 138, at pp. 109, 112–20.

[140] General comment under Article 40 (4) of the ICCPR No. 23/50, adopted on 6 April 1994. Doc. CCPR/C/21/Rev.1/Add.5, reprinted in (1994) 15 *Human Rights Law Journal* 234, 236.

[141] Reprinted in (1995) 34 ILM 351, Article 1(2), 1(3).

[142] United Nations GA Res. 47/135, 18 December 1992, Article 15 (reprinted in [1993] 32 ILM 911).

[143] Florence Benoit-Rohmer and Hilde Hardeman, "The Representation of Minorities in the Parliaments of Central and East European Europe" (1994) 2 Int'l J. Group. Rts 91, 94. In these recent constitutions, there is recognition of the right to participate in public life, but only the new Romanian Constitution guarantees this right, in Article 59, prescribing mandatory representation of at least one seat for each group of citizens belonging to a national minority. Lithuania and Poland have specific provisions that lower the threshold requirements for parties representing minorities. Benoit-Rohmer and Hardeman, *Representation of Minorities* at p. 100.

[144] See the decision of the Bulgarian Constitutional Court No. 1/1992, of 21 April 1992, English summary in (1992) *European Current Law Year Book* 304 (the right of a party representing a minority to take part in the general elections). For background to this case, see Slavi Pashovski, "Minorities in Bulgaria" in John Packer and Kristian Myntti (eds.), *The Protection of Ethnic and Linguistic Minorities in Europe* (Turku, Finland, Institute for Human Rights, Abo Akademi University, 1993), pp. 67, 70–5. In Croatia, see the

Particular emphasis has been placed on the participatory rights of indigenous peoples, whose culture and religion are intimately linked to the natural resources they have been using for generations and are under constant threat from modernization.[145] The World Bank provides a striking example of an institution where the opportunity has been granted to indigenous groups to take part in decisions affecting them.[146]

Subsidiarity

We concluded in chapter 6 that creating links between the regional transnational institution and the sub-state entities, such as provinces or towns, could be a potentially powerful way of overcoming the tension between the institution and the participating national governments. We also noted the relative dearth of attention in international law to this possibility and the corresponding concern that the lack of a legal infrastructure for such undertakings may very well hinder their occurrence. The aim of this section is to shed more light on this relatively dim sphere.

Current doctrine seems to suggest that sub-state agreements will not be governed by international law, but, rather, by one national legal system or a combination of a number of systems. This doctrine derives from two principles: first, the principle of unity of action of the state at the international level; and second, sub-state entities' lack of legal personality in the international sphere.[147] Indeed, it is hardly surprising

Decision of 14 December 1994, summarized in *Bulletin on Const. Case-Law* 223 (1994) (concerning participation in parliamentary voting); Decision of 2 February 1995, summarized in *Bulletin on Constitutional Case-Law* 18 (1995) (the power of a county to determine minority rights). In Romania, see the decision of 18 July 1995, summarized in *Bulletin on Const. Case-Law* 188 (1995) (approving the constitutionality of the proposed education act, which provided proportional representation of professors and teachers from minority groups in the administrative bodies of educational institutions).

[145] See the Draft UN Declaration on the Rights of Indigenous Peoples, adopted by the UN Sub-Commission on Prevention of Discrimination and Protection of Minorities on 26 August 1994 (reprinted in [1995] 34 ILM 541). For examples of the types of conflicts that arise, see *supra*, chapter 5, notes 9–11, 44, 78–9 and accompanying text.

[146] Benedict Kingsbury, "Operational Policies of International Institutions as Part of the Law-Making Process: The World Bank and Indigenous Peoples" in Guy S. Goodwin-Gill and Stefan Talmon (eds.), *The Reality of International Law: Essays in Honour of Ian Brownlie* (Oxford, Clarendon Press, 1999), p. 323.

[147] Maria Teresa Ponte Iglesias, "Les accords conclus par les autorités locales de différents Etats sur l'utilisation des eaux frontalières dans le cadre de la coopération transfrontalière" in (2/1995) *Schweizerische Zeitschrift fuer Internationales und Europaeisches Recht* 103, 122–4; Pierre-Marie Dupuy, "La coopération régionale transfrontalière et

that national governments seek to maintain their monopoly as the sole representatives of their constituencies. As expected, international law, grounded on and developed through state consent, reflects this cartel of power in the hands of national governments. But this doctrine may lead to inefficient outcomes,[148] for it may increase the transaction costs of sub-state negotiators who negotiate with neighboring communities on the other side of the border. These negotiators must select the national law of either one of the parties or of a third party. The assignment of one party's law as the governing law of the agreement may be perceived by the others as domination of the agreement by that party and, therefore, will be resisted by the other parties. More importantly, such assignment exposes the agreement to the possibility of unilateral modifications of the governing law by the state whose law has been chosen and renders it vulnerable to one-sided changes of policy by the central government of that state. It also limits any possible "evolutionary" interpretation of the parties' undertakings in light of the developing norms of international law. Lacking the general rules of international law as background and de-fault rules that could govern such agreements, the sub-state negotiators must reinvent the wheel every time they bargain over the formulation of their agreements. Often they are not even aware of the need to set their agreement in a valid international law framework or else lack the resources necessary for doing so and therefore simply defer and possibly exacerbate the problem.[149]

As the numerous instances have shown, sub-state cooperation is often less costly to establish and maintain than inter-governmental agree-ments are. There is, therefore, good reason for developing rules that can guide the contracting parties and provide authoritative assistance in the event of breach. One possibility is to conclude a framework con-vention that would provide norms to govern such agreements. Just as the 1986 Vienna Convention on the Law of Treaties between States and Inter-national Organizations or between International Organizations[150] came in response to an emerging need to regulate an evolving phenomenon,

---

le droit international" (1977) 23 Ann. Franc. Dr. Int'l. 837, 852; Ulrich Beyerlin, "Transfrontier Cooperation Between Local or Regional Authorities", *Encyclopedia of Public International Law.* (Installment 6), 350.

[148] See Joel P. Trachtman, "L'Etat c'est Nous: Sovereignty, Economic Integration and Subsidiarity" (1992) 33 Harv. Int'l L.J. 459, 472 (on the justification of the existence of political institutions as reducing transaction costs).

[149] See Ulrich Beyerlin, "Transfrontier Cooperation," note 147, at p. 353 (about half of the agreements reviewed lacked any reference to a governing law).

[150] Concluded at Vienna on 21 March 1986, UN Doc. A/CONF.129/15.

a similar convention may be necessary in this context to address the problematics of sub-state agreements. Softer avenues, such as suggested model rules, could also prove helpful in guiding sub-state negotiators.

The Council of Europe has set an important precedent in this respect. In 1980, it adopted a framework convention to facilitate sub-state cooperation.[151] This convention obligates state parties[152] to facilitate and foster such cooperation, to "encourage any initiative by territorial communities or authorities," and to "endeavour to resolve any legal, administrative or technical difficulties likely to hamper the development and the smooth running of transfrontier cooperation."[153] The convention also provides two sets of outline agreements – one set envisioning sub-state arrangements backed by the national governments of the partners and the other set providing agreements solely for the sub-state level – which local representatives may adopt and modify according to their specific requirements. Some of these agreements provide for default choice of law rules, referring to the national law of one of the parties, such as the law of the party whose local authority provides "the principal service, or failing this, the local authority with the most important financial involvement,"[154] or, when a transfrontier private law association is established, the law of the country where the association's headquarters are situated.[155] Other agreement models provide (incomplete) arbitration mechanisms for settling disputes.[156] Further guidelines were prescribed in 1995 in the Additional Protocol to the convention.[157]

[151] European Outline Convention on Transfrontier Co-operation Between Territorial Communities or Authorities, ETS No. 106, Madrid, 21 May 1980, reprinted in http://www.coe.fr/eng/legaltxt/106e.htm.
[152] So far, twenty states have ratified the Convention and eight more recently signed it.
[153] Articles 1, 3, 4 of the Convention, note 151.
[154] See Article 3 of Model Agreement No. 1.4 (Model Inter-State Agreement on Contractual Transfrontier Co-Operation Between Local Authorities).
[155] Model No. 2.11 (Model Agreement on the Creation and Management of Transfrontier Parks Between Private Law Associations, Article 1.2).
[156] See e.g., Model 2.4 (Outline Contract for the Provision of Supplies or Services Between Local Authorities in Frontier Areas), Article 5 (applicable law), and Article 7 (arbitration procedure that does not provide for cases where one of the parties refuses to appoint an arbitrator).
[157] ETS No. 159, adopted on 9 November 1995, entered into force on 1 December 1998, after having been ratified by five states (http://www.coe.fr/tablconv/159t.htm). Three salient principles are put forth by this Protocol:

   (1) States will recognize and respect such agreements provided they conform with national law and with the states' international commitments (Article 1.1). The latter commitments will not be affected by the agreement which entails only the responsibilities of the contracting sub-state parties (Article 1.2);

My suggestion is that a similar effort should be made to formulate international norms that would serve as the background and default norms for sub-state agreements. Until this happens, it may be argued that absent evidence of the parties' express wishes to the contrary, sub-state agreements will be assumed to be governed by international law, both as the objective and subjective laws of the agreements. International law may be regarded as the objective law of such agreements on the basis of the express or implied delegation of powers from the central governments to the lower-level components. Such a delegation of power may be viewed as bestowing upon the sub-state actors, for the purposes of these agreements, the status of a subject of international law, capable of undertaking international obligations.[158] Under this doctrine, the central government would be estopped, both under domestic constitutional law as well as under international law, from setting aside such agreements. In this sense, sub-state undertakings would bind the respective governments vis-à-vis their domestic sub-state entities and vis-à-vis the other governments whose sub-state entities are the other parties to the agreement.

The law on transboundary resources must not remain indifferent to this promise of enhancing local cross-border interaction through sub-state cooperation; it must complement its recognition of the validity of sub-state agreements with their endorsement as a potentially fruitful mode of cooperation. Such a positive signal can be embodied in a principle of subsidiarity similar to the subsidiarity principle under EU law, which allows action "only if and insofar as the objectives of the proposed action cannot be sufficiently achieved by the Member States and can therefore, by reason of the scale or effects of the proposed action, be better achieved by the Community . . ."[159] In this context, it

---

(2) Policies adopted under the agreement will have the status of measures taken under the national legal system of the entities involved (Article 2), and will be subject to the same supervision imposed by the law of each contracting state (Article 6.1);

(3) The applicable law will be the national law of the state in which the headquarters of the joint body are located (Article 4.1 and 6.2).

[158] *Cf.* Christoph Schreuer, "The Waning of the Sovereign State: Towards a New Paradigm for International Law?" (1993) 4 Eur. J. Int'l L., 447, 455–6 ("The logical outcome of these developments would be a general opening up of the treaty process for non-state actors to the extent that they have assumed the functions covered by the respective treaties").

[159] Article 5 (ex Article 3b) of the Consolidated Version of the Treaty Establishing the European Community (reprinted in http://ue.eu.int/Amsterdam/en/traiteco/en1.htm).

is important to emphasize the significance of the 1997 Helsinki Declaration, adopted by the Meeting of the Parties to the 1992 Convention on the Protection and Use of Transboundary Watercourses and International Lakes.[160] Under the Declaration, the parties undertake to "ensure the protection and sustainable use of transboundary waters by cooperating closely at the regional, sub-regional, national, provincial and local levels" and to "delegate relevant activities to the lowest appropriate level." From the perspective of the Watercourses Convention, this entire discussion sounds like science fiction: the Convention is limited strictly to state-to-state interaction.

Conclusion

The clash between the two competing philosophies – what I have called the integrative and the disengagement philosophies – is manifest throughout the discussion of the norms that deal with the various aspects of the negotiation and management of transboundary ecosystem institutions. The question is which philosophy is to prevail. For states belonging to the disengagement camp, the Watercourses Convention offers an attractive conception of water-related agreements as discrete arm's-length agreements and of transnational ecosystem institutions as merely one possibility amongst many at their discretion to consider. In contrast, states pursuing an integrative approach will find support in the letter and spirit of the ICJ decision in the *Gabcikovo–Nagymaros* case and the elaborations presented in this book.

Whereas the Watercourses Convention deliberately refrains from characterizing the type of rights riparians have in "international watercourses," the ICJ decision contains references to the Danube River as a "shared resource," to the notion of a "community of interest" that gives rise to a "common legal right" and to an obligation to further promote common utilization of shared water resources. Whereas the Watercourses Convention confines its purview to the system of surface water and "unconfined" groundwater as constituting "international watercourses," the ICJ decision takes a wider approach, looking also to the environmental impacts of water uses. Whereas the Watercourses Convention gives precedence to existing water-related treaties and insulates them from future developments, the ICJ construes such treaties as subject to evolving norms on environmental protection, which are based

[160] Report of the First Meeting, note 137, at p. 16.

on new scientific findings and new standards set by the international community. Finally, the court embraced two related notions that the Convention rejects: first, the ongoing, rather than discrete, character of water-related agreements, and second, the preference of joint management over litigation as the preferred mechanism in dispute resolution.

What is striking is that the ICJ turns to the Watercourses Convention as the basis of its authority to declare new law, pouring "new law in old bottles."[161] The Convention had barely been signed at the time of the decision and was yet to be ratified by any of the signatory parties.[162] But the court elevated it instantly to the status of customary law, declaring it evidence of "modern development of international law," without offering any further proof of its customary status.[163] But much more important is the fact that the court transformed the meaning of the Convention and co-opted it to serve integrative goals. It is only mere coincidence that the ICJ decision came just four months after the adoption of the Convention by the UN. For these reasons, the *Gabcikovo–Nagymaros* decision "marks a milestone in the evolution of international water law and international environmental law."[164] This milestone is good news for those who are committed to regional cooperation in the use of shared natural resources. The seizing of the opportunity to eclipse the Watercourses Convention was undoubtedly the best thing environmentalists could have hoped for.

But the *Gabcikovo–Nagymaros* decision raises serious doctrinal questions concerning the legitimacy of the court's innovative moves. From where did the ICJ derive its authority to make new law? Why did the ICJ manage to push the law forward, whereas the multilateral negotiation process failed to do so? Indeed, the discussion in this chapter exposes not only the different manifestations of the clash between the philosophies, but also the clash between the two general modalities for the evolution of international law: the progressive development of the law through the

---

[161] A. E. Boyle, "The *Gabcikovo–Nagymaros* Case: New Law in Old Bottles" (1998) 8 Yb. Int'l Envt'l L. 13.

[162] Three years after its adoption, only twelve states had signed the treaty and only half of them had ratified it. *Multilateral Treaties Deposited with the Secretary-General*, Doc. A-51–869 (http://untreaty.un.org/ENGLISH/bible/englishinternetbible/partI/ chapterXXVII/treaty27.asp). This is still only a fraction of the thirty-five ratifications required for the treaty's entry into force.

[163] See note 3, para. 85, p. 147.

[164] Charles B. Bourne, "The Case Concerning the Gabcikovo–Nagymaros Project: An Important Milestone in International Water Law" (1997) 8 Yb. Int'l Envt'l L. 6, 11.

International Law Commission, leading to general conventions on the one hand, and the pronouncements on what the law is by the International Court of Justice. The contrast between the paths taken by the two institutions demonstrates the political limitations that hinder the ILC's work and the prescription of general conventions, from which the ICJ is relatively free. This calls attention to the process of international law-making, the subject of the final chapter.

8 Efficiency, custom, and the evolution of international law on transboundary resources

Introduction

The previous chapter raised a puzzle. From where did the International Court of Justice (ICJ) draw its authority to reshape the law to the extent that it did in the *Gabcikovo–Nagymaros* decision?[1] The court did not employ the traditional method of ascertaining the law, namely, a thorough inspection of state practice to trace the evolution of customary law. Nor did it seek the elusive concept of *opinio juris* to determine which action resulted from the "belief that this practice is rendered obligatory."[2] Instead, the court took a short-cut by invoking the 1997 Convention on the Law of the Non-Navigational Uses of International Watercourses[3] ("the Watercourses Convention") as evidence of the development of modern international law – in other words, a reflection of evolving customary law.[4] This was despite the fact that the Watercourses Convention had been adopted less than four months earlier, had no signatories at the time, and its entry into force was far off.

---

[1] *Gabcikovo–Nagymaros Project (Hungary/Slovakia)*, Judgment, ICJ Reports 1997, p. 7, reprinted in http://www.icj-cij.org/idocket/ihs/ihsjudgement/ihsjudframe1.htm; (1998) 37 ILM 167.

[2] *North Sea Continental Shelf (FRG v. Den./Neth.)*, ICJ Reports 1969, p. 3, at para. 77. The Permanent Court of International Justice first enunciated the doctrine of *opinio juris* in the *Lotus Case (France v. Turkey)*, PCIJ Reports, Series A, No. 10 (1927) at 28: "only if such [practice] were based on their being conscious of having a duty to abstain would it be possible to speak of an international custom." See also *Military and Paramilitary Activities in and against Nicaragua (Nicar. v. US)*, Merits, Judgment, 1986 ICJ Reports, p. 14, at para. 207; *The Paquete Habana*, 175 US 677 (1900).

[3] United Nations Convention on the Law of the Non-Navigational Uses of International Watercourses (adopted on 21 May 21 1997), reprinted in (1997) 36 ILM 700. See *ibid.* for the details of the votes cast.

[4] The *Gabcikovo–Nagymaros* case, note 1, at para. 86.

In September 1997, when the *Gabcikovo–Nagymaros* decision was rendered, state practice and *opinio juris* were rather precarious stilts to serve as the foundation of modern customary watercourse law. But the Watercourses Convention also stood on shaky ground. Despite having been adopted by a strong majority of 103 states, all were yet to sign the Convention. Even more disturbing, two of the dissenters to the Convention were key regional riparians. China, the upper riparian in the Mekong Basin, refused to join its lower riparians on the Mekong River Commission.[5] Turkey, the upper riparian of the Euphrates Basin, refused to negotiate with Syria and Iraq on the sharing of the Euphrates River.[6] The in total twenty-seven abstentions also included important regional states, among them neighboring countries, such as Egypt and Ethiopia, France and Spain, and India and Pakistan, and states involved in regional disputes over water, such as Bolivia,[7] Israel,[8] and Uzbekistan.[9] Obviously, these dissenting and abstaining states balked from committing themselves to regional cooperation in a general, framework convention before entering into direct negotiations over the use of their regional resources.

Nevertheless, the ICJ relied on the Watercourses Convention as the basis of its decision. It is in this context that the puzzle becomes even more complex, for the court distorted the clear logic of the Convention and treated it as an empty vessel into which it injected its own, contradictory vision, pouring "new law in old bottles."[10] In spite of all this, the decision withstood scholarly scrutiny and was received with resounding enthusiasm.[11]

---

[5] See chapter 1, note 54 and accompanying text.

[6] See chapter 1, notes 57, 59 and accompanying text.

[7] On the unresolved dispute between downstream Bolivia and upstream Chile concerning the Chilean diversion of the Lauca River, see Monica Gangas Geisse and Herman Santis Arenas, "Chile–Bolivia Relations: The Lauca River Water Resources" in Gerald H. Blake *et al.* (eds.), *The Peaceful Management of Transboundary Resources* (London, Graham & Trotman, 1995), p. 277.

[8] Eyal Benvenisti and Haim Gvirtzman, "Harnessing International Law to Determine Israeli–Palestinian Water Rights" (1993) 33 Nat. Res. J. 543.

[9] Downstream Uzbekistan has been involved in an ongoing dispute with Tajikistan over the use of the Amu Darya and Syr Darya Rivers; Laurence Boisson de Chazournes, "Elements of Legal Strategy for Managing International Watercourses: The Aral Sea Basin" in Salman M. A. Salman and Laurence Boisson de Chazournes (eds.), *International Watercourses: Enhancing Cooperation and Managing Conflict* (Washington, DC, World Bank, 1998), p. 65.

[10] A. E. Boyle, "The *Gabcikovo–Nagymaros* Case: New Law in Old Bottles" (1998) 8 Yb. Int'l Envt'l L. 13.

[11] Boyle, "*Gabicikovo–Nagymaros* Case" p. 13 See also Charles B. Bourne, "The Case Concerning the Gabcikovo-Nagymaros Project: An Important Milestone in

This chapter responds to this puzzle by arguing that in its decision in the *Gabcikovo–Nagymaros* case, the ICJ was activating its unique legislative role in the international system, the role that empowers it to "leapfrog"[12] over international law. This is the power, under certain conditions, to create new law under the pretext of "finding" existing customary international law. The court has in fact an authority to invent the custom. This chapter argues that states accept this role and welcome such leaps, because they have a general interest in this residual judicial–legislative function. The ICJ can fulfill this function when these leaps produce Pareto-superior norms, provided that at the relevant time, it is the only institution capable of making the leaps. This occurs when high transaction costs prevent states from negotiating bilateral or multilateral agreements.

The Watercourses Convention is but one, recent example of the predicament in which states taking part in multilateral negotiations over a framework agreement refuse to make concessions or even indicate future readiness to offer concessions, because the situation does not ensure reciprocal concessions. In global processes of this kind, states stick to non-cooperative positions in anticipation of the subsequent negotiations over the fate of the transboundary resources they share. The result – in our case, the Watercourses Convention – can only be disappointing. In such circumstances, the ICJ is the sole institution capable of taking the necessary steps towards the development of the law, a sort of trustee acting in the best interests of the states and the global community.

This chapter proposes that when multilateral negotiations fail to reach efficient outcomes or when such negotiations never take place due to conflicting state interests, the international legal system has granted the ICJ the power – and duty – to offer legislative, Pareto-superior remedies. Moreover, the use – or, as the case may be, the abuse – of customary international law is the main vehicle for executing this function. The chapter explores the link between efficiency and the doctrine on customary international law. It analyzes why judicial reliance on custom sometimes fails, suggests why the ICJ is authorized to modify the custom in such cases, and offers the legal base from which the court can inform

International Water Law" (1997) 8 Yb. Int'l Envt'l L. 6 at 11; A. Dan Tarlock, "Safeguarding International River Ecosystems in Times of Scarcity" (2000) 3 U. Denv. Water L. Rev. 231 at 244–7.

[12] David Caron, "The Frog that Wouldn't Leap: The International Law Commission and its Work on International Watercourses" (1992) 3 Colo. J. Int'l Envt'l L. & Pol'y 269.

itself as to the desired content of the new law. My argument is that this legislative function is itself grounded in customary international law.

Customary international law as a proxy for efficiency

The argument developed in this part is that the doctrine on customary international law is inherently linked to the principle of efficiency. Efficiency justifies the doctrine. Put differently, efficiency is the underlying principle – the *Grundnorm* – of customary international law.

An efficient norm in this context is a norm that offers an optimal allocation of global or transnational resources among states. A legal and political environment consisting of sovereign states is one important constraint imposed on the range of possible optimal outcomes: state sovereignty – as it is understood today – entails the authority of states to use resources under their sole ownership at their discretion, even inefficiently.[13] The second constraint is the lack of global mechanisms for the redistribution of welfare among states. The efficient outcome is that which maximizes global welfare given these two constraints. As I argue below, considerations of fairness do not play a role in this context, although in recent years human rights considerations do emerge as an important constraint marginally affecting the range of optimal outcomes.

At the foundation of the argument lies the observation that prevalent and consistent state practice – the necessary component for constituting customary international law – will develop if, and only if, such practice is efficient from the perspectives of most of the governments taking part in the process. Diplomatic immunity and protection of prisoners of war are examples of efficient norms that reinforce themselves through reciprocity, without the need to invoke legal arguments or to resort to adjudication. Because of this affinity between efficiency and prevalent and consistent state practice, lawyers from Grotius onwards have relied on state practice as a proxy for identifying the efficient mode of action states should adopt. Identifying custom has served as a convenient proxy for identifying the efficient norm, especially for lawyers untrained in economics. But this proxy, as with any proxy, is not always an accurate reflection. Market failures often prevent states from adopting a consistent practice or from adapting an existing practice to

---

[13] Nico Schrijver, *Sovereignty over Natural Resources: Balancing Rights and Duties* (Cambridge, Cambridge University Press, 1997).

changed circumstances. My argument, then, is that when this occurs and the proxy fails, lawyers must turn to economic and other tools to discern the efficient norms. I contend that this in fact has been the role played by international adjudicators, especially the ICJ. When they identify a market failure, international courts, and, above all, the ICJ, have the opportunity to step in and try to change the equilibrium[14] of the interstate game or at least try to create conditions that will bring about a change in the equilibrium and thus correct the failure. It is here that they have to invent new law implicitly. I further argue that the international community has recognized such a role as lawful.

When general and persistent state practice takes shape, it moulds itself around rules that the relevant governments find to be efficient. Regarding such rules as legally binding serves the purpose of imposing costs on inactive or weak actors who did not take part in shaping the rules or eliciting compliance from free-riders who seek to deviate from the rules. When a practice becomes inefficient due to changed circumstances, states begin to exert pressure to modify the custom so that it reflects the efficient practice under the new conditions. This is true in respect to the evolution of customary norms among individuals.[15] This is true in respect to the evolution of customary norms among states. As Michael Byers observed, "The customary process operates to maximise the interests of most if not all States by creating rules which protect and promote their common interests;"[16] and this "customary process" reflects the "fact that States generally behave in accordance with their own perceived interests, in so far as they are able to manifest them."[17]

---

[14] A Nash equilibrium is defined as "a steady state of the play of a strategic game in which each player holds the correct expectation about the other players' behavior and acts rationally." Martin J. Osborne and Ariel Rubinstein, *A Course in Game Theory* (Cambridge, MA, MIT Press, 1994), p. 14.

[15] Robert C. Ellickson, *Order without Law: How Neighbors Settle Disputes* (Cambridge, MA, Harvard University Press, 1991), pp. 123–264; Robert D. Cooter, "Decentralized Law for a Complex Economy: The Structural Approach to Adjudicating the New Law Merchant" (1996) 144 U. Pa. L. Rev. 1643. Models of the evolution of the common law towards efficiency are often based upon a bias in litigation in favor of more intensive and extensive challenges to inefficient laws. The first to develop this idea was Paul H. Rubin, "Why Is the Common Law Efficient?" (1977) 6 J. Leg. Stud. 51. *See* also Robert D. Cooter and Lewis Kornhauser, "Can Litigation Improve the Law without the Help of Judges?" (1980) 9 J. Leg. Stud. 139; Eric A. Posner, "Law, Economic, and Inefficient Norms." (1996) 144 U. Pa. L. Rev. 1697.

[16] Michael Byers, *Custom, Power and the Power of Rules* (Cambridge, Cambridge University Press, 1999), p. 19.

[17] Byers, *Custom*, p. 19.

The prime and perhaps oldest examples of efficient state practice yielding to clear customary norms are the laws of war and the laws on diplomatic immunity.

Because general and persistent state practice reflects efficiency, at least from the perspective of state governments, the study of state practice has become a convenient and undisputed proxy for the study of efficiency. It was Hugo Grotius who invented in his treatise *De jure belli ac pacis*[18] the method of observing state practice as a basis for determining the law.[19] Grotius was in fact using state practice to discover efficient norms. At that time, in a world ruled by papal edicts, state practice as such had no legitimacy. Grotius referred to state practice in the classic world because he thought it would make sense for states in the emerging Westphalian order to study and emulate efficient and stable behavior. The Grotian invention caught on not only because it paved the way to "neutral" – i.e., not religiously based – rules, but also because it made sound economic sense: consistent state practice was a good proxy for efficient behavior.

The Grotian enterprise was clearly bent on efficiency. This becomes clear in his first and much celebrated treatise *Mare liberum,* published in 1609,[20] on freedom on the high seas. At that time, this issue had crucial economic implications and therefore proved a bone of contention among European powers. Having little state practice to base his claim upon, Grotius grounded the principle of freedom of navigation directly on considerations of efficiency, using the following analogy:

If any person should prevent any other person from taking fire from his fire or light from his torch, I should accuse him of violating the law of human society, because that is the essence of its very nature...*why then, when it can be done without any prejudice to his own interests, will not one person share with another things which are useful to the recipient, and no loss to the giver?*[21]

This principle of private law, which Grotius drew from Greek sources, was so self-evident to him that he did not find it necessary to explain

[18] Hugo Grotius, *De iure belli ac pacis libri tres. In quibus jus naturae & gentium: item juris publici praecipua explicantur* (1632).

[19] David J. Bederman, "Reception of the Classical Tradition in International Law: Grotius' De jure belli ac pacis" (1996) 10 Emory Int'l L. Rev. 1.

[20] Hugo Grotius, *The Freedom of the Seas*, trans. by Ralph von Deman Magoffin, James Brown Scott (ed.) (Oxford University Press, 1916).

[21] *Ibid.*, p. 38 (emphasis added).

extending its application into the international realm. It is, in Grotius' words, "the law of human society," a *Grundnorm* of the newly asserted law of nations. This is, of course, a definition of efficiency later reformulated by Pareto. Per Grotius, then, international law is based on efficiency and the doctrine on customary law is one of the tools for reflecting efficiency. When technology opens up new horizons for economic and other activities, state practice has yet to crystallize and, hence, cannot serve as a proxy for the efficient norm. Efficiency is then used directly, applied to the particular area of analysis, be it the use of maritime, space, or transboundary resources.

Custom and market failures

Not only when new technology is introduced will customary law not reflect efficiency. Market failures also prevent the development of general and persistent state practice. For reasons discussed in chapter 3, the practice of states, in contrast to the practice of individuals, has a tendency *not* to reflect efficiency.

In the international context, there are two potential sources of market failure: international and transnational. First, there is the international market failure when conflicts of interests among states prevent them from agreeing on certain practices as binding. In the international context, the possibilities for imposing obligations on free-riding actors are much more limited than in the domestic context. The prescription of norms through multilateral or regional treaties is a collective-action problem that often deters states from investing resources in such efforts. Risk aversion leads states to cling to the old rule that ensures them some benefits, rather than risk a reexamination of their entitlements. This is especially true in the case of framework conventions, when participating states are required to forgo their entitlements without having secured the cooperation of their neighbors. Thus, for example, states like India, Turkey, Spain, or France are not inclined to commit themselves to new rules on fresh water or effluent allocations before their neighbors do so or when specific cooperative regimes must still be negotiated at the regional level.

The second, transnational market failure occurs when certain domestic interest groups impose externalities on other domestic interest groups, using international instruments to do so.[22] Because in most

---

[22] On this issue, see the discussion in chapter 3.

states, there are obstacles to giving voice to the larger, less enfranchised domestic groups, state practice will favor the enfranchised minority of the international community. This is particularly so in the sphere of transboundary resource management, where transnational conflicts are abundant and there are more opportunities to capture the negotiating governments.

The first type of market failure will fail to yield general and persistent practice or else fail to modify the practice to accommodate technological and other changes, because some states will refuse to recognize as binding practices that benefit other states. Such disputes will result in conflicting conceptions regarding the binding customary norm. The second type of market failure may yield general and persistent practice only if the groups with similar interests control all governments involved. In such circumstances, state practice will reflect efficient outcomes only for the elite groups and small domestic interest groups that capture their governments.

When market failures prevent the formation of general and persistent state practice, efficiency, as the *Grundnorm*, calls upon adjudicators, primarily the ICJ, to step in and impose an efficient norm. Their decisions have the capability to alter the equilibrium at which states are situated in the game to a more efficient one. The key to the potential contribution of these adjudicators is their readiness to invent customary law. Judges respond to this call to create new law, and their response is subsequently endorsed by the international community. A number of decisions illustrate this point.

The first example relates to the issue of freedom of navigation on a river shared by two states or more. One possible equilibrium is reflected in a norm denying the existence of freedom of navigation and prescribing that each state has the sovereign power to exclude any foreign ship at its discretion. A different equilibrium is reflected in a norm under which all ships enjoy freedom of navigation on the river and its tributaries. The second equilibrium is Pareto-superior to the first because it reduces the costs entailed in interstate commerce. As is often the case, however, states refuse to agree to move from the first inefficient equilibrium to the second, more efficient one. It is at this juncture that judicial intervention can be instrumental. The judgment handed down by the Permanent Court of International Justice ("PCIJ") in the *Case Relating to the Territorial Jurisdiction of the International Commission of the River Oder*[23] is a case in point. The PCIJ had little state practice to rely on in support of

[23] PCIJ, Ser. A, No. 23 (1929).

freedom of navigation. Moreover, its landmark previous decision in the *Lotus* case[24] had implied that the court could impose a duty on states to permit free navigation within their territories only if such a duty is proved to exist under treaty or customary law. Since no proof of any such duty could be shown in the *River Oder Case*, the court opted for the efficient outcome. The court's explanation for its decision was brief. The PCIJ posited that such an outcome is mandated by "the requirements of justice and the considerations of utility."[25] This decision has since been resorted to as proof of the existence of a duty to allow free navigation on international watercourses.[26]

The *Trail Smelter Arbitration* is yet another example of adjudication that nudges states towards a Pareto-superior equilibrium despite a lack of relevant state practice.[27] In this case, the tribunal found Canada in violation of a duty to prevent activities within its territory from causing injury in or to the territory of another state. Absent clear pronouncements of this principle by other international tribunals, the tribunal followed, "by analogy," in the footsteps of three decisions handed down by the US Supreme Court.[28] Although it did not rely explicitly on the efficiency argument, the tribunal did point to the saliency of this factor as part of their reasoning. It asserted that "great progress in the control of fumes has been made by science in the last few years and this progress should be taken into account."[29] Despite meager evidence of state practice to support the decision, the norm prescribed was never questioned. It has since become a cornerstone of international environmental law.[30]

A third example is the arbitration in the matter of *Lac Lanoux*.[31] In this case, the tribunal rejected Spain's contention that France was precluded from making any change concerning the use of a river shared with Spain without the latter's consent. Although the tribunal referred to "international practice" and to customary international law,[32] it did

---

[24] See note 2. On the impact of the "Lotus principle" see chapter 2, text accompanying note 1.

[25] See note 23, at p. 27.

[26] Lucius Caflisch, "Règles générales du droit des cours d'eau internationaux" (1989–VII) 219 *Recueil des Cours* 9 at 32–3, 109–10.

[27] *The Trail Smelter Case (US v. Canada)*, (1949) 3 UN Reports of Arbitral Awards 1905, reprinted in Annual Digest of Public International Law Cases (1938–40), 315.

[28] *Ibid.*, at 318.   [29] *Ibid.*

[30] Phillipe Sands, *Principles of International Environmental Law* (Manchester, Manchester University Press, 1995), p. 191; Patricia W. Birnie and Alan E. Boyle, *International Law and the Environment* (Oxford, Clarendon Press, 1992), pp. 89–90.

[31] *Lac Lanoux* Arbitration, (1957) 24 Int. L. R. 101.

[32] See, for example, *Lac Lanoux* Arbitration, at 130.

not provide any examples of such practice to support its finding. Instead, it emphasized the inefficiency of Spain's assertion of what the tribunal regarded as "a 'right of veto', which at the discretion of one State paralyzes the exercise of territorial jurisdiction of another."[33] Nonetheless, the *Lac Lanoux* decision is hailed as an important milestone in the development of international freshwater law.[34]

The final example explores a slightly different context in which international adjudicators are given the opportunity to intervene in a situation of market failure and set a new, more efficient norm. This is the case of a challenge to the validity of an established customary norm in light of new technology, new scientific findings, or a change in natural conditions that render past practices inefficient and require the adoption of new, more efficient rules. Such developments create a time lag between what is established as customary law and the more efficient behavior dictated by the new reality. At such junctures, a wedge is created between efficiency and custom, as well as corresponding pressure to amend the law. Absent market failures, such pressure ultimately leads to efficient outcomes. But when market failures prevent such legal modifications, judges are given the opportunity to intervene. Such an opportunity seemed to present itself to the ICJ judges who presided over the *Fisheries Jurisdiction* case.[35]

Reacting to overfishing by British and German fishing fleets in the North Sea, in 1972, Iceland extended its Exclusive Fisheries Zone from a twelve- to fifty-mile limit. This move was in clear violation of the old customary norm of freedom on the high seas. But it was in line with the demands of efficiency and sustainability: it was aimed at preventing a tragedy of the commons due to overfishing. Iceland had been preoccupied with this possible tragedy since 1948, when the Althing passed the Law concerning the Scientific Conservation of the Continental Shelf Fisheries.[36] Furthermore, the Althing's resolution in 1972 provided, *inter alia*, that "effective supervision of the fish stocks in the Iceland area be continued in consultation with marine biologists and that the necessary measures be taken for the protection of the fish stocks and specific areas in order to prevent over-fishing."[37] The extension of the zone of unilateral

---

[33] *Ibid.*, at 128.
[34] Sands, *Principles*, note 30, at pp. 348–9; Birnie and Boyle, *International Law*, note 30, at pp. 102–3.
[35] *Fisheries Jurisdiction Case* (*United Kingdom* v. *Iceland*), ICJ Reports 1974, p. 3. See also the parallel and almost identical Fisheries Jurisdiction Case (*Federal Republic of Germany* v. *Iceland*), ICJ Reports 1974, p. 175.
[36] *Fisheries Jurisdiction Case* (*United Kingdom* v. *Iceland*), note 35, para 19.
[37] *Ibid.*, para. 29.

appropriation resulted in most of the fisheries in the North Sea becoming the private property of Iceland, the coastal state. Commanding sole authority over the exclusive zone, Iceland could now manage alone the harvests and thereby prevent depletion and ensure sustainable yields. This outcome was in the long-term interests also of states with large fishing fleets who had already started to compete among themselves and overfish. Approval of Iceland's unilateral measure was mandated on efficiency grounds. But it was contrary to prevalent practice at the time. The ICJ, however, refused to leapfrog the law and side with efficiency and sustainability. Instead, it resorted to the traditional search for past practice accepted as law. Coming as no surprise, in examining past practice, the court could detect no customary norm that allowed coastal states to extend their exclusive spheres of economic interest beyond the twelve-mile territorial sea zone. Iceland, and the efficiency principle, lost. At that very stage, however, the law was transforming rapidly, as the court itself indicated in subsequent judgments. Between 1976 and 1979, about two-thirds of the exclusive economic zones and exclusive fishery zones of up to two hundred miles had been unilaterally created,[38] with the relevant states choosing not to wait for the results of the ongoing negotiations of the UN Convention on the Law of the Sea. This practice enabled the ICJ, a decade after its *Fisheries Jurisdiction* decision, to rule that a custom had emerged in support of the legality of an exclusive economic zone of two hundred miles.[39]

The reason for the decision in the *Fisheries Jurisdiction* case was not the court's disregard for the principle of efficiency. Rather, the judges chose to defer to the governments that were negotiating the Third Conference on the Law of the Sea. They explicitly acknowledged that their legislative role is residual and becomes relevant only after states have failed to come to an agreement:

The very fact of convening the third conference on the Law of the Sea evidences a manifest desire on the part of all States to proceed to the codification of that law on a universal basis, including the question of fisheries ... Such a general desire is understandable since the rules of international maritime law have been the product of mutual accommodation, reasonableness and co-operation. In the circumstances, the Court, as a court of law, cannot render judgment *sub specie legis ferendae*, or anticipate the law before the legislator has laid it down.[40]

---

[38] R. R. Churchill and A. V. Lowe, *The Law of the Sea* (Manchester, Manchester University Press, 1988), pp. 144–6.

[39] *Continental Shelf (Libyan Arab Jamahiriya/Malta)*, Judgment, ICJ Reports 1985, p. 13, para. 34.

[40] *Fisheries Jurisdiction* case (*United Kingdom* v. *Iceland*), note 35, at para. 53.

Iceland's extension of its exclusive fisheries zone did, indeed, deviate from the prevailing practice at the time; but, I submit, this was not unlawful. This measure was a quintessential first step towards the establishment of a new norm through new state practice, the first challenge to the old law, heralding the birth of the new law. Iceland's unilateral act, like the assignment of property rights to continental shelves, was an exemplary case of how science and technology postulate the need to change international law.[41] It was, therefore, commensurate with the abstract, basic norm of efficiency. Nevertheless, in view of the fact that multilateral negotiations on a new law were under way, there was no immediate need for judicial intervention. As it turned out, the international market in fact proved efficient, and no judicial intervention was ultimately required.

These examples, like the 1997 *Gabcikovo–Nagymaros* decision,[42] suggest that international courts and tribunals often decide to take the leap and declare new law. They take this responsibility in light of the adverse environmental and health consequences of the continuation of the prevailing practice of states, provided no contemporaneous negotiations render their intervention unnecessary.

These international adjudicators also have to pretend that they are not doing what they actually are doing. They conceal the new law in "old bottles."[43] They play down their legislative role and instead stress conformity with their duty to apply "international custom, as evidence of a general practice accepted as law."[44] Despite the fact that at the time they are taken, these are novel moves, leaps that are consistent with efficiency, concerning issues like freedom of navigation or the duty to prevent environmental harm, they are likely to be accepted as reflecting the law and to produce consistent future practice. The aftermath of these decisions and the fact that the decisions are subsequently widely and undisputedly accepted as valid pronouncements of international law despite their weak doctrinal basis suggest that states accept the role of tribunals as legislators of Pareto-superior norms. Put differently, these

---

[41] H. Scott Gordon, "The Economic Theory of a Common-Property Resource: The Fishery" (1954) 62 J. Pol. Econ. 124; Colin W. Clark, "Restricted Access to Common-Property Fishery Resources: A Game-Theoretic Analysis" in Pan-Tai Liu (ed.), *Dynamic Optimization and Mathematical Economics* (New York, Plenum Press, 1980), p. 117. See also Yoram Barzel's analysis of the conversion of the North Sea into owned property, Yoram Barzel, *Economic Analysis of Property Rights* (2nd edn, Cambridge, Cambridge University Press, 1997), pp. 101–2.

[42] See note 1.   [43] Boyle, "Gabcikovo–Nagymaros," note 10.

[44] Article 38(1)(b) of the Statute of the International Court of Justice.

cases suggest the existence of a custom that tribunals are authorized to prescribe efficient norms when market failures preclude states from reaching such norms directly.

To conclude, when states or any other players interact, they rationally choose among different equilibria. This may lead them to converge in inefficient equilibria. A judicial declaration of one equilibrium as the one that is binding as custom is likely to lead all players to modify their activities to conform with the judicially sanctioned equilibrium. This equilibrium will thus become the new practice, the new custom. Because this is likely to occur, but more importantly because this new norm is efficient, judges – at the ICJ or elsewhere – are fully authorized to divert from current practices and "detect" the new custom.

One important caveat to this argument concerns the courts', especially the ICJ's, own institutional limitations. Similar to domestic courts, international tribunals are concerned with their own reputation. They have an acute sense of the limits on their enforcement and managerial powers, and they do not wish to produce judgments that will remain unheeded and ineffective. In addition, the ICJ and other tribunals whose jurisdiction is optional are concerned with losing business as a result of susbsequent refusal to litigate. These constraints shape the range of strategies available to the tribunals. When they perceive themselves capable of resolving disputes, they will prefer to assign a liability rule in order to protect entitlements,[45] such as in the case of maritime delimitation. In such contexts they are less concerned with reputational losses, because when the court establishes a more efficient equilibrium as legally binding – such as, for example, new norms on maritime delimitation between adjacent coastal states – this has the potential of quickly generating wide and stable practice and hence increasing global welfare as well as the tribunal's own reputation. In contrast, when the judicial contribution can only be more modest the tendency would be to set property rules to protect states' entitlements, thereby requiring those states to negotiate a solution among themselves, perhaps following a set of broad guidelines declared by the tribunal. These judges will not, for example, set up institutions for collective action for the litigants, but instead they will declare that the litigants should do so themselves through bilateral negotiations.[46]

---

[45] Guido Calabresi and A. Douglas Melamed, "Property Rules, Liability Rules, and Inalienability: One View of the Cathedral," (1972) 85 Harv. L. Rev. 1089.

[46] As was the outcome in the *Gabcikovo–Nagymaros* case, note 1.

Judicial findings on efficiency

The above discussion suggests that a state's deviation from past practice in favor of a more efficient practice need not be regarded as a breach, because the deviation conforms with the underlying norm of efficiency and because such a deviation has a very good chance of becoming, rather soon, the new practice. When the proxy fails, judges have the authority to explore the efficient rule. They need not wait – often in vain – for persistent and general practice to form. Direct reliance on scientific evidence in prescribing efficient and sustainable norms can be the basis for forward-looking decisions.

To find the efficient norm, our judge can turn to the relevant data and its scientific evaluation to study the possible normative responses. This raises evidentiary issues, but such issues are not foreign to the adjudication process, which often involves expert evidence. The litigants or the tribunals themselves can appoint expert witnesses and even assistants to the tribunal. In the *Trail Smelter* case, for example, the two sides appointed scientists – one from each side – to assist the tribunal.[47] When undisputed scientific findings show that a new norm can promise more efficient outcomes, outcomes states cannot reach due to market failures, then it is up to the judges to base their decisions on science as the neutral and efficient norm. Turning to science replaces reliance on the indirect and failed proxy, the custom. Science is the direct, objective guideline to efficient norms. The study of economics and game theory, sociology and psychology, replaces the study of historical facts in the intellectual quest to establish a neutral foundation for international law.

Thus, for example, this science-based theory on the sources of international law suggests that if economic efficiency postulates effective joint management of shared ecosystems, international law should endorse it and fashion norms that will reduce the transaction costs involved in negotiating and establishing such cooperative regimes. Because states balk at conceding to this postulate without eliciting concessions from their neighbors, state practice will often prove to be inefficient. The equilibrium is reluctance to cooperate. Judges and other third parties can overcome this impasse. This observation explains the shortcomings of the 1997 Watercourses Convention and, in contrast, the landmark decision in *Gabcikovo–Nagymaros*.

In fact, the ICJ seems to have implied that it has done exactly this. The link between science, sustainable development, and the law is captured

---

[47] See note 27, at p. 318.

best in the following passage:

Owing to new scientific insights and to a growing awareness of the risks for mankind – for present and future generations – of pursuit of such interventions at an unconsidered and unabated pace, new norms and standards have been developed, set forth in a great number of instruments during the last two decades. Such new norms have to be taken into consideration, and such new standards given proper weight . . . not only when States contemplate new activities but also when continuing with activities begun in the past. This need to reconcile economic development with protection of the environment is aptly expressed in the concept of sustainable development.[48]

In the same vein, the ICJ should use the same science-based approach to shape and mould other legal issues. One example, related to transboundary resources, involves transboundary gas and oil deposits straddling land or maritime boundaries. It is beyond dispute that so-called "unitization" – namely, the formation of a single authority to exploit such deposits – is a prerequisite for efficient utilization.[49] Such a single authority will be able to exploit the underground pressures to propel captured deposits to the surface to benefit all riparians. Absent a joint development agreement among the co-owning states, only a fraction of the deposit can be exploited. Does this suggest that states sharing such resources have a duty under international law to form such joint regimes? If one follows the doctrine on customary law, the answer will be negative. Indeed, a recent examination of state practice typically concluded with the following observation: "[The] survey of bilateral state practice indicates, as a preliminary conclusion, that a rule of customary international law requiring cooperation specifically with a view toward joint development or transboundary unitization of a common hydrocarbon deposit has not yet crystallized."[50] Although such a conclusion may be a fair description of an existing situation, it cannot be shared by judges or arbitrators who have the power and duty to modify the law in the face of market failures. They cannot let pass the opportunity to transform the existing equilibrium of the interstate game into a new and self-enforcing equilibrium of cooperation. A hint in that direction was recently given in the second phase of the Arbitral Award in

---

[48] See note 1, para. 140.

[49] Gary D. Libecap, *Contracting for Property Rights* (Cambridge, Cambridge University Press, 1989); Rainer Lagoni, "Oil and Gas Deposits across National Frontiers" (1979) 73 AJIL 215 at 224.

[50] David M. Ong, "Joint Development of Common Offshore Oil and Gas Deposits: 'Mere' State Practice or Customary International Law?" (1999) 93 AJIL 771 at 792.

the *Eritrea–Yemen Arbitration*.[51] In the process of delimiting the maritime boundary between the two states, the tribunal considered the possibility that petroleum deposits would be found to straddle the boundary. Although it admitted that so far no general customary law has been developed to require unitization, it carefully tailored a specific norm pertaining only to the two litigants. It found that the parties are

> bound to inform one another and to consult one another on any oil and gas and other mineral resources that may be discovered that straddle the single maritime boundary between them or that lie in their immediate vicinity. Moreover, the historical connections between the peoples concerned, and the friendly relations of the Parties that have been restored since the Tribunal's rendering of its Award on Sovereignty, together with the body of State practice in the exploitation of resources that straddle maritime boundaries, import that Eritrea and Yemen should give every consideration to the shared or joint or unitised exploitation of any such resources.[52]

Recourse to science, of course, is not a panacea. As mentioned in chapter 5,[53] science cannot eliminate uncertainties concerning the potential risks of managing shared resources, Thus, science does not relieve decision-makers from their discretion in the adoption of policies. International adjudicators are less capable in making such choices for the litigants and thus tend to let the states negotiate these choices directly. Adjudicators can only require the litigants to negotiate the establishments of joint management institutions, as the most efficient way to resolve commons issues, be it fisheries, forests, fresh water, or oil. Science, in this context the collective-action theory, can serve as the basis for the duty to treat international commons as jointly owned and impose duties on co-owners to establish joint management mechanisms to ensure public participation and other procedural guarantees. This theory may, in principle, eventually be proven wrong. At that stage, a new theory will suggest a more efficient norm that will then become the new norm.

Efficiency and the contemporary crisis of customary law

Emphasis on efficiency, lying at the heart of the customary process, is capable of salvaging the doctrine from several recent challenges to

---

[51] *Eritrea–Yemen Arbitration* (Award Phase II: Maritime Delimitaiton), 17 December 1999 http://www.pca-cpa.org/ERYE2.
[52] *Ibid.*, at para 86.
[53] See the discussion in chapter 5, note 16 and accompanying text.

its authority and utility. To begin with, the view of customary law as a proxy for efficiency resolves the paradox concerning the process of evolution of custom.[54] If customary law evolves and is modified through unilateral action, how does the doctrine explain the first deviation from the custom: Is it an illegal defection? Does subsequent practice absolve the deviating state from responsibility? If a tribunal is faced with a complaint against the deviating state at a crucial point of transition, must the tribunal find against the harbinger of the new law and, thereby, arrest the development of the law? Grounding the doctrine of customary international law on efficiency offers a solution to the seeming paradox: if the deviating state asserts a new norm that is more efficient than the old one, the deviation should not be considered a violation. Thus, in the example of the *Fisheries Jurisdiction* cases,[55] the ICJ was in a position to accept Iceland's unilateral move as legal, prior practice notwithstanding. As I argued, the ICJ should have done so, and its refusal to do so can only be justified in light of the simultaneous efforts to negotiate this matter.

The emphasis on efficiency also assists in addressing the growing criticism from leading scholars, who have come to disparage the bankruptcy of the traditional doctrine in customary international law. Robert Jennings warned of the inconsistency between doctrine and reality, suggesting that "Perhaps it is time to face squarely the fact that the orthodox tests of custom – practice and *opinio juris* – are often not only inadequate but even irrelevant for the identification of much new law today."[56] Michael Reisman warned against resorting to custom as a tool for clarifying and implementing policies in an advanced and complicated civilization.[57] The particular discrepancy between state practice and appropriate norms in the sphere of the environment has been noted by Oscar Schachter, who wrote, "To say that a state has no right to injure the environment of another seems quixotic in the face of the great variety of transborder environmental harms that occur every day."[58] Other commentators have conceded that the list of customary norms

---

[54] On this paradox, see Byers, *Custom*, note 16, at pp. 130–3.

[55] See note 35.

[56] Robert Y. Jennings, "The Identification of International Law" in Bin Cheng (ed.), *International Law: Teaching and Practice* (London, Stevens, 1982), p. 3 at p. 5.

[57] See W. Michael Reisman, "The Cult of Custom in the Late 20th Century" (1987) 17 Cal. W. Int'l L. J. 133 at 134.

[58] Oscar Schachter, "The Emergence of International Environmental Law" (1991) 44 J. Int'l Aff. 457 at 462–3.

concerning the environment is rather short[59] and that the actual identified norms lack precision: "their legal status, their meaning, and the
consequences of their application to the facts of a particular case or
activity remain open."[60]

The doctrine on customary law does, indeed, fail if its role is to be
understood only as the provider of positive norms, because on many
important questions, customary law fails to yield norms based on general and persistent state practice. This predicament tempts scholars to
employ a less rigid scrutiny of state behavior, peppered with value judgments. This practice is as ancient as its inventor, Grotius.[61] But breaking loose from the doctrinal examination of state practice exposes the
judge to the opposite danger: the danger of subjectivity and, hence,
loss of legitimacy.[62] Attempts to reconcile conflicting practice through
normative arguments deliver a blow to the Grotian effort to establish
a value-neutral basis of the law. In a divided globe, this is a source of
crisis for the doctrine. The search for customary law requires judges and
scholars to choose between rigid past-looking, inefficient neutrality and
teleological forward-looking subjectivity.

Basing the doctrine of customary international law on efficiency re-
deems it from yet another fatal flaw. Presented as a neutral doctrine,
the only normative basis the doctrine on customary international law
can have is state consent: custom reflects the express or implied consent
of states to be bound by it. But clearly, state consent can no longer provide the normative basis, the *Grundnorm*, for international obligations.
State consent is no more a satisfying normative basis than the idea of
positivism. A global system redefined as subject to basic principles of human rights cannot be described as preserving the unfettered discretion

---

[59] Sands, *Principles*, note 30, lists the following norms as customary: the responsibility of
states not to cause environmental damage and the principle of good neighborliness
and international cooperation. These are the only principles sufficiently established to
give rise to customary obligation. Birnie and Boyle, *International Law*, note 30, at
pp. 92–4, suggest that the customary duty of states is to take adequate steps (due
diligence) to control and regulate sources of serious global environmental pollution
or transboundary harm within their territories or subject to their jurisdictions.

[60] Sands, *Principles*, note 30, at p. 236.

[61] Bederman, "Reception," note 19, at pp. 37–9: "Grotius assigned varying significance to
natural law dictates and to customary international law evidences in his consideration
of different international law doctrines . . . He purported to scientifically approach the
historical record of State practice, although he was prepared to modify (and even
distort) that evidence in order to fashion rules of enduring significance to modern
nations. Grotius thus embodied the contemporary ambivalence of legal scholarship."

[62] Jonathan I. Charney, "Universal International Law" (1993) 87 AJIL 529 at 545–6.

of states to accept or decline the evolution of the law in conformity with the basic norms. As noted by Martti Koskenniemi,[63] if we postulate about sovereignty, as we do when thinking about national systems, that states have no inherent value, only an instrumental value in ensuring self-determination and allocating competences regarding people and resources, then state consent, in itself, cannot have any normative value. In other words, the seemingly positive, neutral basis of the doctrine not only yields indeterminate outcomes, but also is not satisfactory from a normative perspective. Basing the doctrine on efficiency relieves the need to reconcile custom with consent. It provides an alternative, neutral ground for determining "custom."

Efficiency, equity and fairness: contradiction or affinity?

So far, I have argued that efficiency is the underlying principle of customary international law. This argument serves as the basis for the claim that the process of defining and redefining customary international law could be based on a study of the efficient rules. But aside from efficiency, there is another, even more visible, principle of international law: the principle of equity. The prevalence of equity in different legal contexts of international law may suggest that an underlying policy of equity-as-fairness permeates quite a number of international norms, constituting "an important, redeeming aspect of the international legal system"[64] that may clash and take precedence over efficiency. Moreover, the interplay between equity-as-fairness and efficiency raises the question as to whether in addition to moving international law to the "Pareto frontier," namely, to the zone of the most efficient outcomes, tribunals can also choose among the efficient outcomes the outcome that distributes the gains most equitably. These questions call for an inquiry into the relationship between the two principles: equity-as-fairness and efficiency.

A scientific inquiry into efficient behavior can be challenged by fairness arguments. Fairness arguments may call for two distinct outcomes. First, where a tradeoff between efficiency and fairness is possible, emphasis on fairness may lead to the adoption of policies that are less

---

[63] See Martti Koskenniemi's critique of the principle of state consent in Martti Koskenniemi, *From Apology to Utopia* (Helsinki, Finnish Lawyers' Pub. Co., 1989), pp. 270–3.

[64] Thomas M. Franck, *Fairness in International Law and Institutions* (Oxford, Clarendon Press, 1995), p. 79.

efficient, but distribute more equitably the benefits and risks across the relevant groups. Another, different situation involves no such trade-off, but, instead, situations in which fairness considerations are called upon to decide among two or more options that are equally efficient (namely, that are on the "Pareto frontier"). Distributive considerations inform both situations. In national legal systems, we tend to prefer the second outcome, not because we eschew fairness, but because we can design legal or market-based schemes to redistribute risks and benefits among citizens. Thus, we achieve fair results without sacrificing efficiency. But in the international system, such a solution is problematic because there are no readily available institutions for making and implementing decisions on distributing or redistributing among states the added benefits from the more efficient policies. International tribunals have no criteria or tools to make choices on the Pareto-frontier. Because there are no institutional means to allocate the gains from efficient outcomes, if we stick to fairness we might stop short of reaching efficient outcomes. We may face the tough choice between efficiency and fairness.

My observation is that international law does not make efficiency secondary to fairness arguments. International law recognizes not a doctrine of fairness, but a doctrine of equity, which is quite distinct from fairness. Equity is not "equity-as-fairness." In the international legal context, the concept of "equity" serves an entirely different function from that served by the notion of fairness. Therefore, the pervasiveness of equity tells little about the status of fairness as a principle of international law. In fact, I contend that the concept of equity in international law serves the goal of efficiency. Hence, there is a convergence, rather than a clash, between the two concepts. A careful analysis of the concept of equity supports the observation that efficiency considerations stand at the basis of international law.

The gist of my argument is that the concept of equity in international law does not serve distributive functions. Instead, equity in the application of the law (as distinct from equity *ex aequo et bono*, namely, equity rendered outside the law)[65] serves two functions, both of which are mandated by efficiency. First, equity grants discretion to decision-makers where existing norms are too crude to be applied to specific matters, such as in cases dealing with the delimitation of maritime

---

[65] On the different contexts of equity, see Ruth Lapidoth, "Equity in International Law" (1987) 22 Israel L. Rev. 161.

boundaries. Second, equity creates incentives for users of transboundary resources to act efficiently by cooperating with their neighbors.

Equity as discretion

Just as the doctrine of administrative discretion provides authority to administrative agencies to implement statutory policies in specific instances, so the doctrine of equity allows negotiating states to seek ways to resolve their differences within the confines of international norms that provide only rough principles. More importantly, it authorizes judges or other third parties to balance all the considerations which international law prescribes as relevant to the intricacies of the particular case.[66] "'Equité' peut être définie comme la solution qui convient le mieux à chaque cas qui se présente. Elle est donc autre chose que l' 'Equity' du Droit anglo-saxon."[67] In other words, equity provides decision-makers with the discretion to implement policies on an ad hoc basis. Its use does not guarantee outcomes that are "fair," if by "fair" we mean at least some attention is paid to distributive effects.

This function is evident in the areas of territorial and maritime boundary delimitation. In the delimitation of territorial boundaries in the de-colonized world, equity considerations have played only a marginal role. The reigning principle is *uti possidetis juris*, namely, the supremacy of pre-independence boundaries.[68] This principle promotes stability and certainty and, hence, is efficient. It eschews fairness considerations, even if a most precious resource is kept only on one side of the border. Equity considerations become relevant only when the *uti possidetis* rule fails to provide a clear answer.[69] And even then, equity allows decision-makers the discretion to weigh a host of natural factors, none having to do with distributive concerns.[70]

---

[66] *Continental Shelf (Tunisia/Libyan Arab Jamahiriya)*, Judgment, ICJ Reports 1982, p. 18, at para. 24 (Jimenez de Arechaga, J., sep. op.). See also Masahiro Miyoshi, *Considerations of Equity in the Settlement of Territorial and Boundary Disputes* (Dordrecht, M. Nijhoff, 1993), p. 173.

[67] A. Alvarez, "Preliminary Communication" (1937) 40 Ann. Inst. Dr. Int'l 151.

[68] Miyoshi, *Considerations*, note 66, at pp. 153–4.

[69] *Frontier Dispute (Burkina Faso/Mali)*, Judgment, ICJ Reports, 1986, p. 554, para. 149 ("to resort to the concept of equity in order to modify an established frontier would be quite unjustified").

[70] *Land, Island and Maritime Frontier Dispute (El Salvador/Honduras: Nicaragua intervening)*, ICJ Reports 1992, p. 351, at para. 58 ("economic considerations of this kind could not be taken into account for the delimitation of continental shelf areas . . . still less can they be relevant for the determination of a land frontier").

The doctrine on equity is also effective in settling questions of maritime delimitation of continental shelves or exclusive economic zones between two or more neighboring states with opposite or adjacent coasts. Here, again, equity grants discretion to judges and authorizes them to balance all the relevant conflicting factors and interests.[71] The judges do not necessarily use their discretion to achieve fair outcomes. If we examine the many decisions rendered by the ICJ, by its Chambers, and by other tribunals, we will see that the major consideration has been the geography of the particular area. In one such decision, the ICJ rejected the relevance of the comparable wealth factor. It ruled out the possibility that the relative economic wealth of the two litigants would influence its decision "in such a way that the area of continental shelf regarded as appertaining to the less rich of the two States would be somewhat increased in order to compensate for its inferiority in economic resources," explaining that

such considerations are totally unrelated to the underlying intention of the applicable rules of international law. It is clear that neither the rules determining the validity of legal entitlement to the continental shelf, nor those concerning delimitation between neighbouring countries, leave room for any considerations of economic development of the States in question.[72]

Geographic considerations prevailed because the continental shelf and the exclusive maritime economic zones in question were viewed as the natural prolongation of the landmass. The ICJ explained the extension of jurisdiction over them on the basis of a conceptual nexus between

---

[71] See, e.g., *North Sea Continental Shelf Cases*, note 2, at p. 3, para. 93; *Continental Shelf* (*Tunisia/Libyan Arab Jamahiriya*), note 63, para. 107; Application for Revision and Interpretation of the Judgment of 24 February 1982 in the *Case Concerning the Continental Shelf* (*Tunisia v. Libyan Arab Jamahiriya*), Judgment, ICJ Reports 1985, p. 192, at para. 35. On the role of equity in balancing different factors in the context of maritime delimitation, see generally Francisco Orrego Vicuña, *The Exclusive Economic Zone* (Cambridge, Cambridge University Press, 1989), pp. 211–22; Prosper Weil, *Perspectives du droit de la délimitation maritime* (Paris, Pedone, 1988) pp. 282–5; Malcolm D. Evans, *Relevant Circumstances and Maritime Delimitation* (Oxford, Clarendon Press, 1989), pp. 90–4; Miyoshi, *Considerations*, note 66.

[72] *Tunisia v. Libyan Arab Jamahiriya* 1985 case, note 71, at para. 50. See also *Tunisia v. Libyan Arab Jamahiriya* 1982 case, note 71, at para. 107; Louis F. E. Goldie, "Reconciling Values of Distributive Equity and Management Efficiency in the International Commons" in René-Jean Dupuy (ed.), *The Settlement of Disputes on the New Natural Resources* (The Hague, M. Nijhoff, 1983), p. 335 at pp. 338–9; L. D. M. Nelson, "The Roles of Equity in the Delimitation of Maritime Boundaries" (1990) 84 AJIL 837; Derek W. Bowett, "The Economic Factor in Maritime Delimitation Cases" in *International Law at the Time of Its Codification: Essays in Honour of Roberto Ago*, vol. II (Milan, Giuffre, 1987), p. 45 at pp. 61–2.

the land – sovereignty over land being the basis for the claim – and the shelf or the exclusive economic zone.[73] As the ICJ stated, "Since the land is the legal source of power which a State may exercise over territorial extensions to seaward, it must first be clearly established what features do in fact constitute such extensions."[74]

In using sovereignty over land as the starting point for exercising their discretion, judges disregard the relative economic conditions in the relevant countries. This has been most clearly manifested in the case concerning the delimitation of the continental shelf between Libya and Malta.[75] The ICJ emphasized that equity entails

the principle that there is to be no question of refashioning geography, or compensating for the inequalities of nature; ... the principle that ... [equity does not] seek to make equal what nature has made unequal; and the principle that there can be no question of distributive justice.[76]

Thus, the considerable difference in the economic strength of Libya and Malta was regarded as an irrelevant consideration.[77] A Chamber of the ICJ did consider the economic interests of communities residing within the disputed area (i.e., the local population that relies on the fisheries for subsistence); but it made it clear that these interests would not be assigned great significance and would influence the decision only marginally:

What the Chamber would regard as a legitimate scruple lies rather in concern lest the overall result ... should unexpectedly be revealed as radically inequitable, that is to say, as likely to entail catastrophic repercussions for the livelihood and economic well-being of the population of the countries concerned.[78]

---

[73] *North Sea Continental Shelf Case*, note 2, paras. 19, 96. See also *Aegean Sea Continental Shelf Case* (*Greece* v. *Turkey*), ICJ Reports 1978, p. 3, at para. 86; Research Center for International Law, University of Cambridge, *International Boundary Cases: The Continental Shelf* (Cambridge, Grotius, 1992), vol. I, at p. 12; Elihu Lauterpacht, *Aspects of the Administration of International Justice* (Cambridge, Grotius, 1991), pp. 124–30; Weil, *Perspectives*, note 71, at pp. 56–61; Evans, *Relevant Circumstances*, note 71, at pp. 99–103. Sovereignty over the land as the source of title over the territorial waters was confirmed by the ICJ in the *Anglo-Norwegian Fisheries Case* (*United Kingdom* v. *Norway*), ICJ Reports 1951, at 116, 133.

[74] *North Sea Continental Shelf Case*, note 2, at para. 96.

[75] See Churchill and Lowe, *Law of the Sea*, pp. 144–6.   [76] *Ibid.*, para. 46.

[77] *Ibid.*, para. 50. See also Evans, *Relevant Circumstances*, note 71, at p. 186, and the decision of the ICJ in *Maritime Delimitation in the Area between Greenland and Jan Mayen* (*Denmark* v. *Norway*), ICJ Reports 1993, p. 38, at paras. 79–80.

[78] *Delimitation of the Maritime Boundary in the Gulf of Maine Area* (*Canada/United States of America*), ICJ Reports 1984, pp. 246, at para. 237; Evans, *Relevant Circumstances*, note 71, at pp. 189, 200; Weil, *Perspective*, note 71, at pp. 274–80.

The concept of equity as fairness could have been invoked in the sphere of delimitation of contiguous river boundaries. Equal access to navigable courses could have constituted a very cogent principle. It is, therefore, rather telling that the Beagle Channel arbitration mentions it as the *last* consideration guiding its decision, giving precedence to geographic considerations. According to the tribunal, it was guided "in particular by mixed factors of appurtenance, coastal configuration, equidistance, and also of convenience, navigability, and the desirability of enabling each Party so far as possible to navigate in its own water."[79]

Equity as an efficient incentive

The second function of equity is demonstrated in the sphere of allocation of freshwater resources. It is in this context that the claim for equity as fairness is the most pronounced. But I will argue here as well that equity serves the objective goal of efficiency. The Watercourses Convention sets forth as its goal "attaining optimal and sustainable utilization [of international watercourses] and benefits therefrom."[80] This reflects a long-standing conception, in the words of the Institut de droit international, "that the maximum utilization of available natural resources is a matter of common interest," as well as the aspiration to "assur[e] the greatest advantage to all concerned."[81] But this law does not neglect divergent economic conditions among riparians.[82] Thus, the Helsinki Rules emphasize that states are bound by "a duty of efficiency *which is commensurate with their financial resources*"[83] in the quest for achieving "maximum benefit to each basin State from the uses of the waters with the minimum detriment to each."[84] As the International Law Association explained,

---

[79] *Beagle Channel Arbitration (Argentina v. Chile)* (1977) 52 Int'l L.R. 93, para. 110. See also the *Case Relating to the Territorial Jurisdiction of the International Commission of the River Oder*, PCIJ, Ser. A, No. 23, at 27–8 (1929); see also Eli Lauterpacht, "River Boundaries: Legal Aspects of the Shatt-Al-Arab Frontier" (1960) 9 ICLQ 216–22.

[80] Article 5(1) of the Watercourses Convention.

[81] See the Institute of International Law's Resolution on the Utilization of Non-Maritime International Waters (Except for Navigation) adopted at its session at Salzburg (3–12 Sep. 1961), ((1961) 49 (II) Ann. Inst. Dr. Int'l 370) (translated in 56 AJIL 737 (1962)), Preamble and Article 6.

[82] But for the claim that "only intentional inefficiency" matters, see Ximena Fuentes, "The Criteria for the Equitable Utilization of International Rivers" (1996) 67 Brit. Yb. Int'l L. 337 at 385.

[83] *Commentary on the Helsinki Rules*, (1967) ILA Report of the Fifty-Second Conference, 484, 487 (emphasis added).

[84] *Ibid.*, at 486.

State A, an economically advanced and prosperous State which utilizes the inundation method of irrigation, might be required to develop a more efficient and less wasteful system forthwith, while State B, an underdeveloped State using the same method might be permitted additional time to obtain the means to make the required improvements.[85]

Furthermore, the report of the International Law Commission explicitly emphasizes that,

Attaining optimal utilization and benefits does not mean achieving the "maximum" use, the most technologically efficient use, or the most monetary valuable use... Nor does it imply that the State capable of making the most efficient use... should have a superior claim to the use thereof. Rather, it implies attaining maximum possible benefits for all watercourse States and achieving the greatest possible satisfaction of all their needs, while minimizing the detriment to, or unmet needs of, each.[86]

This concern with "equity of needs" is reflected in the list of factors mentioned as relevant in the process of determining what constitutes "reasonable and equitable" allocation.[87] Included among these factors are "the social and economic needs of the watercourse States concerned"[88] and of "the population dependent on the watercourse in each watercourse State."[89] Although these factors are preceded by "geographic, hydrographic, hydrological, climatic, ecological and other factors of a natural character,"[90] they are not overshadowed by them. It is generally accepted that the natural factors provide only the factual basis for the analysis of the respective needs.[91]

---

[85] *Ibid.*, at 487.

[86] "ILC Report on the Law of the Non-Navigational Uses of International Watercourses" (1994) 2 *ILC Yearbook* (Part 2), at 85, 97. See also *Commentary on the Helsinki Rules,* note 80, at p. 487: "A 'beneficial use' need not be the most productive use to which the water may be put, nor need it utilize the most efficient methods known in order to avoid waste and insure maximum utilisation."

[87] On the "equitable and reasonable" standard, see the discussion in chapter 7, notes 19–43 and accompanying text.

[88] Article 6(1)(b) of the Watercourses Convention. A similar consideration appears in Article V(2)(e) of the 1966 Helsinki Rules, note 83.

[89] Article 6(1)(c) of the Watercourses Convention. A similar consideration appears in Article V(2)(f) of the 1966 Helsinki Rules, note 83.

[90] Article 6(2)(a) of the Watercourses Convention. The Helsinki Rules specify these natural factors as the first three on the list; Article V (2)(a)–(c).

[91] See Bonaya Adhi Godana, *Africa's Shared Water Resources* (London, F. Pinter, 1985), p. 58: "Factors (a) to (c) mentioned in Article V of the Helsinki Rules merely re-emphasise the need for an accurate assessment of the nature and extent of the interdependence between utilisation in the different basin states."

Equity as "equity of needs" is well entrenched in the practice related to federal or international fresh water.[92] In fact, there exists no evidence to support the contrary proposition, namely, that waters should be allocated, for example, according to the contribution of each state to the basin's waters or according to the length of the river in each state's territory.[93] It is interesting to note that the priority of human needs over natural parameters was recognized even by ardent enemies – Israel and its Arab neighbors – during the Johnston negotiations in the 1950s concerning the allocation of the waters in the Jordan River basin. The key for calculating entitlements was the existing and potential irrigated agriculture within the riparian states, rather than each state's natural contribution.[94]

My argument is that the recourse to the "equity of needs" principle in this context is motivated first and foremost by efficiency considerations. "Equity of needs" analysis creates efficient incentives for users of the resource, promotes negotiation between states, instead of litigation, as a means of settling disputes, and increases the likelihood that the negotiated settlement will be domestically ratified and obeyed. In other words, invocation of "equity of needs" in international practice has not – or not only – been an outcome of fairness, but, rather, primarily the outcome of efforts to achieve efficiency. The "reasonable and equitable utilization" principle is an efficient norm.

---

[92] This observation has never been contested. See, e.g., Gerhard Hafner, "The Optimum Utilization Principle and the Non-Navigational Uses of Drainage Basins" (1993) 45 Austrian J. Publ. Int'l Law 113 at 124–6; Patricia Buirette, "Genèse d'un droit fluvial international général" (1991) 95 Rev. Gen. Dr. Int'l Pub., 5 at 38; Gunther Handl, "The Principle of 'Equitable Use' as Applied to Internationally Shared Natural Resources: Its Role in Resolving Potential International Disputes over Transfrontier Pollution" (1978) 14 Rev. Belge Dr. Int'l 40 at 46, 52–4; Jerome Lipper, "Equitable Utilization," in A. H. Garretson, R. D. Haydon, and C. J. Olmstead (eds.), The Law of International Drainage Basins (Dobbs Ferry, NY, New York University School of Law [by] Oceana Publications, 1967), p. 16 at pp. 41, 45; Charles Bourne, "The Right to Utilize the Waters of International Rivers" (1965) 3 Can. Yb. Int'l L. 187 at 199; William Griffin, "The Use of Waters of International Drainage Basins under Customary International Law" (1959) 53 AJIL 50 at 78–9.

[93] Lipper, "Equitable Utilization," note 92, at p. 44: "Factors unrelated to the availability and use of waters are irrelevant and should not be considered. For example, the size of a particular state in relation to a co-riparian or the fact that the river flows for a greater distance through one state than another is not in itself a factor to be considered in determining what is an equitable utilization (although it may prove relevant on the issue of 'need')."

[94] The main factor used to calculate the allotments to the two principal users, Israel and Jordan, was their potential irrigable land. Kathryn B. Doherty, "Jordan Waters Conflict" in International Conciliation No. 553 (New York, Carnegie Endowment for International Peace, 1965), at pp. 25–8.

"Equity of needs" creates the proper incentives for users to invest in efficient uses of a shared watercourse, because efficient existing uses enjoy qualified status in the analysis of the riparians' equitable shares. As Article 8(1) of the Helsinki Rules states, "an existing reasonable use may continue in operation unless the factors justifying its continuance are outweighed by other factors leading to the conclusion that it be modified or terminated so as to accommodate a competing incompatible use."[95] This principle is also accepted in the jurisprudence of federal states, including the US,[96] Argentina,[97] and India.[98] Although the Watercourses Convention seems to depart from that principle,[99] this is yet to reflect actual practice. As explained in the commentary to the Helsinki Rules, "failure to give any weight to existing uses can only serve to inhibit river development. A State is unlikely to invest large sums of money in the construction of a dam if it has no assurances of being afforded some legal protection for the use over an extended period of time."[100] But only efficient projects deserve protection. The principle of optimal utilization – that protects only "beneficial uses"[101] – implies that existing uses that are wasteful do not merit continued respect.[102] Thus, existing

[95] The 1966 Helsinki Rules, note 83, Article 8(1). In the same vein, see the IDI's Salzburg Resolution, note 81, Articles 3, 4.

[96] The last ruling of the US Supreme Court on the subject, in *Colorado* v. *New Mexico* (459 US 176, 103 S. Ct. 539, 74 L.Ed.2d 348 [1982]), emphasized the predominance of existing uses and placed a heavy burden on the state challenging such uses to prove the desirability of the proposed change. The Court stated, "We recognize that the equities supporting the protection of existing economies will usually be compelling." *Colorado* v. *New Mexico* at 187. To justify the detriment in existing uses, a state would have to "demonstrate by clear and convincing evidence that the benefits of the [change] substantially outweigh the harm that might result." *Ibid.* See also R. Simms, "Equitable Apportionment – Priorities and New Uses" (1989) 29 Nat. Res. J. 549; A. Dan Tarlock, "The Law of Equitable Apportionment Revisited, Updated, and Restated" (1985) 56 U. Col. L. Rev. 381.

[97] *Province of La Pampa* v. *Province of Mendoza*, Argentina Supreme Court of Justice, December 1987 (reported in *International Rivers and Lakes* No. 10, May 1988, at 2). On this case, see Fuentes, *Criteria*, note 82, at pp. 382–5.

[98] The Rau Commission decision (mentioned in Lipper, *Equitable Utilization*, note 92, at 51).

[99] Article 6(1)(e) of the Convention couples "existing and potential uses" and further on stipulates that neither of the factors enjoys precedence (Article 6[3]). See also chapter 7, notes 118–19 and accompanying text.

[100] *Commentary on the Helsinki Rules*, note 83, at 493.

[101] As Lipper defines this term, "a use, to be entitled to protection, must afford sufficient economic and social benefit to the user so that it is reasonable, under all the circumstances, that its continuation be considered" (*Equitable Utilization*, note 92, at pp. 63.)

[102] *Colorado* v. *New Mexico*, note 96 (a more efficient future use may outweigh an existing wasteful one); Lipper, *Equitable Utilization*, note 92, at p. 46 (a more efficient use by

uses enjoy priority,[103] "but a contemplated use will nevertheless prevail over an existing use if the former offers benefits of such magnitude as is sufficient to outweigh the injury to the existing use."[104]

"Equity of needs" is also efficient for the creation of the "constructive ambiguity" of the legal norm that is so important for creating the proper incentives for states to commence negotiations. As discussed at length in chapter 7,[105] a vague standard will prompt riparians to negotiate, whereas a clearer rule will provide an incentive to litigate (if rights have been or are likely to be infringed) or to refuse to negotiate towards regional cooperation. A vague standard that instructs states to provide information not only on the natural characteristics of the shared transboundary resource but also on their existing and potential needs, provides incentive to both negotiate and bring pertinent information to the negotiation table. A discussion over existing and potential needs sensitizes negotiators to the constraints of their partners and the limitations on their room for political maneuvering and enables them to explore ways to accommodate the interests of all parties.

Finally, "equity of needs" raises the potential of domestic support for negotiated or judicial allocation of entitlements. Domestic users, especially the strong agricultural interest groups, will simply resist new allocations that severely curtail their existing uses. Moreover, it would be much more difficult to implement reallocation plans. The conditional priority assigned to existing uses assists in reducing domestic opposition to the ratification and implementation of agreements.

For these reasons, the "equity of needs" principle does not contradict efficiency. On the contrary, it facilitates efficiency. This result is in line with the other manifestations of equity in international law. Equity in international law is first and foremost an efficient norm. In many cases it has nothing to do with fairness. Even if it turns out to be commensurate with fairness, this is not its main rationale and is a product of coincidence alone. Fairness can and will become relevant only when adjudicators choose between various possibilities for the allocation of resources on the "Pareto frontier" – namely, without compromising efficiency.

---

another state is not dispositive, but it is a relevant consideration). Yet, there are examples in interstate settings in which inefficient uses were permitted to continue; see, e.g., *La Pampa* v. *Mendoza*, note 95; the Indus (Rau) Commission's decision, India, 1942, note 98, at p. 47.

[103] Lipper, *Equitable Utilization*, note 92, at p. 58. See also Caflisch, *Règles générales*, note 26, at pp. 158–60.

[104] Lipper, *Equitable Utilization*, note 92, at p. 58.

[105] See chapter 7, notes 16–23 and accompanying text.

Efficiency, human rights, and self-determination

A real question of tradeoff exists when efficiency clashes with human rights and self-determination considerations. It is here that efficiency yields. A state has no duty to allow other states to utilize a resource under its sole ownership and may decide not to utilize it at all or to utilize it inefficiently. At this juncture efficiency yields to the principle of self-determination, which entails a state having the power to exercise sovereignty over its natural resources.[106]

Similarly, efficiency does not prevail over human rights concerns. As elaborated in chapter 5,[107] the human rights perspective generates a host of principles concerning the management of transboundary resources. It implies mandatory provision of sufficient allocation on a per capita basis of basic resources that have a direct impact on human welfare. In particular, it mandates a sufficient supply of clean air and water for personal consumption for all individuals, regardless of nationality, financial resources, or other distinguishing factors. It requires minimum and equally distributed exposure to risks. It entails the protection of minority groups – their property and culture – against government-sponsored development projects that disregard them. At these junctures, efficiency is subordinated to basic human rights considerations. There can be no tradeoff, for example, between water for basic domestic use and water for irrigated cash crops. In the same vein, the unequal distribution of risks of pollutants among different regions or groups of people infringes on the principle of equal treatment of individuals. Damming rivers or diverting flows from one basin to another may increase the availability of water for some people, but, at the same time, create adverse environmental and social effects for others. In such cases, equality requires a careful balancing between the interests of the different communities and fair representation of the affected groups in the various stages of the decision-making process.

These constraints are not only self-evident from the human rights perspective; they are also instrumental. Granting voice and paying respect to individual and communal interests and rights enhances the quality of the decisions that take due account of their concerns, increases the legitimacy of such agreements, and, thus, strengthens the durability and success of collective action.[108] Thus, in a deeper sense, there is no

---

[106] See chapter 1, notes 48–9 and accompanying text; Schrijver, *Sovereignty over Natural Resources*, note 13.

[107] See chapter 5, notes 57–70 and accompanying text.

[108] See discussion in chapter 6, notes 23–4 and accompanying text.

conflict between efficiency and human rights. Under conditions of growing scarcity, of recurring crises and natural disasters, the law of human rights postulates sustainability as a goal of international law. In the context of transboundary resource management, states are required to pursue policies that provide efficient and sustainable uses.

Concluding observations

Efficiency has been, all along, the driving force behind the development of international law in general and customary international law in particular. State practice has often proven a reliable proxy for determining what constitutes efficient behavior for all states to follow. This has enabled international tribunals and other actors to impose sanctions on free-riders seeking to deviate from the efficient norm. But this proxy fails when global or regional conditions lead states to pursue inefficient behavior. In such situations, tribunals and other third parties can make a difference by pushing states towards new, more efficient equilibria. The argument developed in this chapter is that the judicial authority to nudge states towards efficient equilibria exists under international law. It derives from the principle of efficiency that nurtures much of international law. Where state practice fails to follow the efficient mode of behavior, there is recourse to scientifically based analysis. In the sphere of transboundary management, the study of collective action and of the characteristics of the transboundary resource in question can suggest norms of behavior that are efficient. Judges in international tribunals, especially at the International Court of Justice, therefore have a unique role in the advancement of international law. They have the genuine opportunity to translate science into law, an opportunity the states themselves often fail to seize. In a sense, tribunals have the opportunity to declare as law what states would have agreed to had they decided behind a Rawlsian veil of ignorance under the assurance of reciprocity.[109] This explains why in the sphere of shared fresh water, judicial solutions such as in the *Gabcikovo–Nagymaros* case have offered far greater promise and reached far more efficient prescriptions than internationally negotiated framework conventions.

This analysis explains why the decisions of the International Court of Justice are so central to the international legal system. Despite the fact

---

[109] John Rawls, *A Theory of Justice* (Cambridge, MA, Belknap Press of Harvard University Press, 1971).

that ICJ decisions are technically not binding on states that have not taken part in the specific litigation and despite the fact that the court is not bound by its prior decisions, states rarely dispute its decisions and are quick to modify their expectations accordingly. The extraordinary weight accorded to these decisions only serves to highlight the underlying first-order preference of all states: to have an institution that can overcome market failures and prescribe norms they are otherwise unable to agree upon.

This analysis further demonstrates the need to redirect the focus in the study of international law. The analysis of international law cannot remain confined to the study of past precedents. In order to remain true to the underlying goals of international law, it must encompass the scientific insights in the fields related to the subject-matter under scrutiny. This observation ultimately justifies the approach taken in this book.

9   Conclusion

This book's first objective is to outline conditions for collective action in the management of transboundary resources, especially transboundary ecosystems. For that purpose, the book analyzes how states shape their preferences in the transnational competition over resources. Observation of state practice leads to the second goal of the book: to explain why states fail to act together with their neighbors to protect their shared resources, and suggests how these failures could be corrected. This goal leads to the third step in the inquiry: to examine what responses international law has to offer for inducing states to opt for cooperation. The divergence within contemporary international law, particularly the clash between two conflicting approaches – "the philosophy of disengagement" and "the philosophy of integration" – calls for a final insight: international tribunals, particularly the International Court of Justice (ICJ), are authorized to develop international law in accord with the goal of efficiency.

In exploring the potential for collective state action, this book breaks away from the distinction between domestic and international processes. The systemic failures of states that derive from their heterogeneity in their management of natural resources affect similarly domestic and international resources. Domestic conflicts of interest – between farmers and city dwellers, between producers and consumers – are responsible for distorted governmental choices, whether on the domestic or the international level. In fact, many domestic interest groups cooperate with *foreign* interest groups in order to impose their externalities on their respective rival *domestic* groups. The better-organized and, hence, more politically effective domestic interest groups – usually producers, employers, and service-suppliers – cooperate with their counterparts in different states to exploit collectively the less organized groups in those states, such as the consumers, the employees, and the environmentally affected citizens. Thus, many global collective action failures must be attributed to conflicts between warring domestic groups rather than to international competition.

Hence solutions must be found both in the domestic and the international components of the governmental decision-making process. Therefore, this book adopted a view of the management of shared natural resources as a transnational challenge. Rejecting the still-prevailing Westphalian perception of nation states as unitary actors, this book preferred a different paradigm – the transnational conflict paradigm – that explains better various collective-action failures and provides guidance with regard to feasible mechanisms for correcting these failures. At its core lies the observation that states are not monolithic entities and many of the pervasive conflicts of interest are in fact more internal than external and stem from the heterogeneity within states.

This paradigm points to the necessary framework for transnational cooperation. In a sense, the paradigm offers a blueprint for the legal engineers who would – indeed, should – be called upon, once the political will is forged, to design the institutions for cooperation. Having identified the domestic roots of many conflicts over shared resources, I examine whether states – or rather governments representing states in the international sphere – could be made to reflect more equitably and democratically the long-term interests of their respective constituencies and to what extent international law and institutions could be instrumental in such a scheme. I claim that such an endeavor is possible. In order to promote efficient, just, and sustainable allocation of transboundary resources through collective action, it is necessary to set up mechanisms that will ensure all relevant actors a voice in decision-making, monitoring, and enforcement processes, just as was the case in the ancient "village republics." To do so, the current juxtaposition of seemingly highly regulated national procedures and quite flexible and even obscure international procedures must be eradicated and replaced. This analysis of existing institutions and norms at the national and international level sets the stage for the search for the legal and institutional remedies that would level the political playing field and increase the potential of the long-term perspective.

Ecosystem management requires a constant balancing of conflicting interests and even human rights, under constraints imposed by nature and by the limited ability of humans to assess risks. To meet these challenges, it is necessary to structure the decision-making processes

in transboundary ecosystem institutions in ways that will ensure informed and unbiased decisions. Such institutions could sustain markets
for some of the uses of the resource, say, for trade in pollution permits,
for water for irrigation, or for reclaimed sewage water intended for agriculture, provided the institution can ensure the attainment of the tasks
of resource-, claims-, and risk-management and can comply with the
external normative constraints.

Chapter 6 outlines the modalities for the operation of transnational
ecosystem management institutions. It suggests that institutions that
follow the principles identified in this chapter are likely to adopt policies that are efficient, equitable, and sustainable and reflect a fair and
balanced weighing of the interests of all communities sharing the common resource or resources. In its footsteps, chapter 7 discusses the status
of contemporary international law in this area, examining whether or
not it corresponds to the demands set forth in the previous chapter.
Chapter 7 discovers a clash between what I have called the integrative and the disengagement philosophies. The question is which philosophy prevails. For states belonging to the disengagement camp, the
1997 Watercourses Convention offers an attractive conception of water-
related agreements as discrete arm's-length agreements and of transnational ecosystem institutions as merely one possibility amongst many at
their discretion to consider. In contrast, states pursuing an integrative
approach will find support in the letter and spirit of the ICJ decision
in the *Gabcikovo–Nagymaros* case and the elaborations presented in this
book.

Chapter 7 revealed the ironic use by the ICJ of the 1997 Watercourses
Convention as a tool to declare the *opposite* approach as having the status
of customary international law. The court transformed the meaning of
the Convention and co-opted it to serve integrative goals. This decision, I
argued in chapter 8, is consistent with the role of the ICJ in developing
and updating international norms in order to obtain efficient use of
resources, through declaring new norms as having the status of custom
or by interpreting bilateral and multilateral treaties flexibly. In short, the
customary process is a prescriptive process aiming at reaching efficient
state behavior. International judges are the agents of the international
community authorized to point the direction of legal development.

We have come to the end of the journey. The question is whether the
integrative philosophy will prevail over the disengagement philosophy
in the long run. Some even worry about the growing propensity of some
states to resort to force to secure shares in key transboundary resources

such as fresh water. I belong to the camp of the cautious optimists who are inclined to adopt the view that as demands for fresh water and the environment intensify and diversify and supplies dwindle, interdependency becomes greater. I believe that transnational collective action in the utilization of transboundary resources can, in principle, provide optimal and sustainable results. A bleak future of wars over control of water resources is not an unavoidable tragedy in our new millennium. Growing interdependency increases the incentive of neighboring states to cooperate in formulating entitlements and norms and in managing shared resources. Thus, in the near future, as supplies of finite transboundary resources in general fail to meet rising demands, more and more states may be unable to afford the luxury of ensuring their relative power gap through unilateral appropriation. They will sooner or later opt for making the most from the collective-action situation.

Bibliography

Abbott, Kenneth W. and Duncan Snidal, "Why States Act through Formal International Organizations" (1998) 42(1) J. Conflict Res. 3.

Ackerman, Bruce A., "Beyond Carolene Products" (1985) 98 Harv. L. Rev. 713.

Ackerman, Bruce A., and David Golove, "Is NAFTA Constitutional?" (1995) 108 Harv. L. Rev. 799.

Adalsteinsson, Ragnar, "The Current Situation of Human Rights in Iceland" (1994) 61/62 Nordic J. Int'l L. 167.

Adams, William M., *Wasting the Rain: Rivers, People and Planning in Africa* (London, Earthscan, 1992).

Agora: "May the President Violate Customary International Law?" (1986) 80 AJIL 913.

Agora: "What Obligations Does Our Generation Owe to the Next? An Approach to Global Environmental Responsibility" (1990) 84 AJIL 190.

Alhértière, Dominique, "Settlement of Public International Disputes on Shared Resources: Elements of a Comparative Study of International Instruments" (1985) 25 Nat. Res. J. 701.

Allan, J. A., "Overall Perspectives on Countries and Regions" in Peter Rogers and Peter Lydon (eds.), *Water in the Arab World: Perspectives and Prognoses* (Cambridge, MA, Harvard University Press, 1994) p. 65.

Alston, Philip, "International Law and the Human Right to Food" in Philip Alston and Katarina Tomasevski (eds.), *The Right to Food* (Boston, M. Nijhoff, 1984), p. 9.

Alvarez, A., "Preliminary Communication" (1937) 40 Ann. Inst. Dr. Int'l., 151.

Anaya, S. James, "Maya Aboriginal Land and Resource Rights and the Conflict Over Logging in Southern Belize" (1998) 1 Yale H. R. & Dev. L.J. 17.

Anbar, Ali Hasan Dawod, "Socio-Economic Aspects of the East Ghor Canal Project" (Ph.D. thesis, University of Southampton 1983).

Anderson, Terry L. and Pamela Snyder, *Water Markets: Priming the Invisible Pump* (Washington, DC, Cato Institute, 1997).

236

Applegate, John S., "Beyond the Usual Suspects: The Use of Citizens Advisory Boards in Environmental Decisionmaking" (1998) 73 Ind. L.J. 903.

"A Beginning and not an End in Itself: The Role of Risk Assessment in Environmental Decision-Making" (1995) 63 U. Cin. L. Rev. 1643.

Armstrong, Neal E., "Anticipatory Transboundary Water Needs and Issues in the Mexico–US Border Region in the Rio Grande Basin" (1982) 22 Nat. Res. J. 877.

Arrow, Kenneth J., *Social Choice and Individual Values* (New York, Wiley, 1951).

Arsanjani, Mahnoush H., "The Rome Statute of the International Criminal Court" (1999) 93 AJIL 22.

Auer, Matthew R., "Domestic Politics and Environmental Diplomacy: A Case from the Baltic Sea Region" (1998) 11 Georgetown Int'l Envtl L. Rev. 77.

Axelrod, Robert, *The Evolution of Cooperation* (New York, Basic Books, 1984).

Bagley, Edgar S., "Water Rights Law and Public Policies Relating to Ground Water 'Mining' in the Southwestern States" (1961) 4 J. Law & Econ. 144.

Baird, Douglas G., Robert H. Gertner, and Randal C. Picker, *Game Theory and the Law* (Cambridge, MA, Harvard University Press, 1994).

Barberis, Julio A., "The Development of International Law of Transboundary Groundwater" (1991) 31 Nat. Res. J. 167.

"Bilan de recherches de la section de langue française du Centre d'étude et de recherche de l'Académie" in Centre for Studies and Research, *Rights and Duties of Riparian States of International Rivers* (Dordrecht, M. Nijhoff, 1990), p. 15.

Barzel, Yoram, *Economic Analysis of Property Rights* (2nd edn, Cambridge, Cambridge University Press, 1997).

Bauer, Carl J., "Bringing Water Markets Down to Earth: The Political Economy of Water Rights in Chile, 1976–95" (1997) 25 *World Development* 639.

Baxter, Richard R., "The Indus Basin" in Albert H. Garretson, Robert D. Hayton, and Cecil J. Olmstead (eds.), *The Law of International Drainage Basins* (Dobbs Ferry, NY, Institute of International Law, New York University School of Law, Oceana Publications, 1967), p. 443.

Bayefsky, Anne and Joan Fitzpatrick, "International Human Rights Law in United States Courts: A Comparative Perspective" (1992) 14 Mich. J. Int'l L. 1.

Beaumont, Peter, "The Qanat: A Means of Water Provision from Groundwater Sources" in Peter Beaumont, Michael Bonnie, and Keith McLachlan (eds.), *Qantas, Kariz and Khattara: Traditional Water Systems in the Middle East and North Africa* (London, Middle East & North African Studies Press, 1989), p. 13.

Becker, Gary S., "A Theory of Competition among Pressure Groups for Political Influence" (1983) 98 Q. J. Econ. 371.

Bederman, David J., "Reception of the Classical Tradition in International Law: Grotius' De jure belli ac pacis" (1996) 10 Emory Int'l L. Rev. 1.

Bekker, Peter H. F., *The Legal Position of Intergovernmental Organizations* (Dordrecht, M. Nijhoff, 1994).

Benoit-Rohmer, Florence and Hilde Hardeman, "The Representation of Minorities in the Parliaments of Central and East European Europe" (1994) 2 Int'l J. Group. Rts. 91.

Benvenisti, Eyal, "Exit and Voice in the Age of Globalization" (1999) 98 Mich. L. Rev. 167.

"Margin of Appreciation, Consensus and Universal Standards" (1999) 31 NYU J. Int'l L. & Pol. 843.

"National Courts and the International Law on Minority Rights" (1997) 2 ARIEL 1.

"Collective Action in the Utilization of Shared Water Resources: The Challenges of International Water Resources Law" (1996) 90 AJIL 384.

"The Influence of International Human Rights in Israel: Present and Future" (1995) 28 Israel L. Rev. 136.

"Judges and Foreign Affairs: A Comment on the Resolution of the Institute of International Law on 'National Courts and the International Relations of their State'" (1994) 5 Eur. J. Int'l L. 423.

"Judicial Misgivings Regarding the Application of International Norms: An Analysis of Attitudes of National Courts" (1993) 4 Eur. J. Int'l L. 159.

"The Israeli–Palestinian Declaration of Principles: A Framework for Future Settlement," (1993) 4 Eur. J. Int'l L. 542.

"The Applicability of Human Rights Conventions to Israel and to the Occupied Territories" (1992) 26 Israel L. Rev. 24.

Benvenisti, Eyal and Haim Gvirtzman, "Harnessing International Law to Determine Israeli–Palestinian Water Rights" (1993) 33 Nat. Res. J. 543.

Berber, Friedrich J., *Rivers in International Law* (London, Stevens, 1959).

Bermann, George A., "Constitutional Implications of US Participation in Regional Integration" (1998) 46 *American Journal of Comparative Law* 463.

Beschorner, Natasha, *Water and Instability in the Middle East*, Adelphi Paper 273 (London, Brassey's, 1992).

Bessette, Joseph M., *The Mild Voice of Reason: Deliberative Democracy and American National Government* (University of Chicago Press, 1994).

Beyerlin, Ulrich, "Transfrontier Cooperation between Local or Regional Authorities," Encyclopedia of Public Int'l L. (Installment 6) 350.

Birnie, Patricia W. and Alan E. Boyle, *International Law and the Environment* (Oxford, Clarendon Press, 1992).

Blatter, Joachim and Helen Ingram, "States, Markets and Beyond: Governance of Transboundary Water Resources" (2000) 40 Nat. Res. J. 439.

Blum, Jonathan, "The Deep Freeze: Torts, Choice of Law, and the Antarctic Treaty System: Is It Adequate to Regulate or Eliminate the Environmental Exploitation of the Globe's Last Wilderness?" (1992) 14 Hous. J. Int'l L. 597.

Bodansky, Daniel, "Scientific Uncertainty and the Precautionary Principle" (Sept. 1991) 33 *Environment* 4.

Bogdan, Michael, "Application of Public International Law by Swedish Courts" (1994) 63 Nordic J. Int'l L. 3.

du Bois, François, "Social Justice and the Judicial Enforcement of Environmental Rights and Duties" in Alan E. Boyle and Michael R. Anderson (eds.), *Human Rights Approaches to Environmental Protection* (Oxford, Clarendon Press 1996), p. 153.

Boisson de Chazournes, Laurence, "Elements of a Legal Strategy for Managing International Watercourses: The Aral Sea Basin" in Salman M. A. Salman and Laurence Boisson de Chazournes (eds.), *International Watercourses: Enhancing Cooperation and Managing Conflict, Proceedings of a World Bank Seminar* (Washington, DC, World Bank, 1998), p. 65.

Bourne, Charles B., "The Case Concerning the Gabcikovo–Nagymaros Project: An Important Milestone in International Water Law" (1997) 8 Yb. Int'l Envnt'l L. 6.

"The International Law Commission's Draft Articles on the Law of International Watercourses: Principles and Planned Measures" (1992) 3 Colo. J. Int'l Env. L. & Pol'y 65.

"Procedure in the Development of International Drainage Basins: The Duty to Consult and to Negotiate" (1972) 10 Can. Yb. Int'l L. 212.

"Mediation, Conciliation and Adjudication in the Settlement of International Drainage Basin Disputes" (1971) 9 Can. Yb. Int'l L. 114.

"The Right to Utilize the Waters of International Rivers" (1965) 3 Can. Yb. Int'l L. 187.

Bouwer, H., "Reuse of Water: A Sustainable Perspective" in J. C. Van Dam and J. Wessel (eds.), *Transboundary River Basin Management and Sustainable Development* (1993), vol. I, p. 89.

Bowett, Derek W., "The Economic Factor in Maritime Delimitation Cases", in *International Law at the Time of Its Codification: Essays in Honour of Roberto Ago*, vol. II (Milan, Giuffre, 1987), p. 45.

Boyd, Jamie W., "Canada's Position Regarding an Emerging International Fresh Water Market with Respect to the North American Free Trade Agreement" (1999) 5 NAFTA L. & Bus. Rev. Am. 325.

Boyle, Alan E., "The *Gabcikovo–Nagymaros* Case: New Law in Old Bottles'" (1998) 8 Yb. Int'l Envt'l L. 13.

"The Role of International Human Rights Law in the Protection of the Environment" in Alan E. Boyle and Michael R. Anderson (eds.), *Human Rights Approaches to Environmental Protection* (Oxford, Clarendon Press, 1996), p. 43.

Bradley, Curtis A. and Jack L. Goldsmith, "Customary International Law as
    Federal Common Law: A Critique of the Modern Position" (1997) 110
    Harv. L. Rev. 815.
  "The Current Illegitimacy of International Human Rights Litigation" (1997)
    66 Fordham L. Rev. 319.
Bramble, Barbara J. and Gareth Porter, "Non-Governmental Organizations and
    the Making of US International Environmental Policy" in Andrew Hurrell
    and Benedict Kingsbury (eds.), *The International Politics of the Environment*
    (Oxford, Clarendon Press, 1992), p. 313.
Breckenridge, Lee P., "Nonprofit Environmental Organizations and the
    Restructuring of Institutions for Ecosystem Management" (1999) 25
    Ecology L.Q. 692.
Breyer, Stephen, *Breaking the Vicious Circle: Toward Effective Risk Regulation*
    (Cambridge, MA, Harvard University Press, 1993).
Briggs, Herbert W., *The International Law Commission* (Ithaca, NY, Cornell
    University Press, 1965).
Brill, Eyal, "Applicability and Efficiency of Market Mechanisms for Allocation of
    Water with Bargaining" (Ph.D. dissertation, submitted to the Senate of The
    Hebrew University of Jerusalem, 1997, in Hebrew).
Brilmayer, Lea, "International Law in American Courts: A Modest Proposal"
    (1991) 100 Yale L.J. 2277.
Bronars, Kris and Sarah Michaels, Annotated Bibliography on Partnerships for
    Natural Resource Management (1997), available at
    http://www.icls.harvard.edu/ppp/contents.htm#sources
Bruhacs, Janos, *The Law of Non-Navigational Uses of International Watercourses*
    (Dordrecht, M Nijhoff 1993).
Brunnée, Jutta and Stephen J. Toope, "Freshwater Regimes: The Mandate of the
    International Joint Commission" (1998) 15 Ariz. J. Int'l & Comp. Law 273.
  "Environmental Security and Freshwater Resources: Ecosystem Regime
    Building" (1997) 91 AJIL 26.
  "Environmental Security and Freshwater Resources: A Case for International
    Ecosystem Law" (1994) 5 Yb. Int'l Envt'l L. 41.
Buergenthal, Thomas, "To Respect and Ensure: State Obligations and
    Permissible Derogations" in Louis Henkin (ed.), *The International Bill of
    Rights: The Covenant on Civil and Political Rights* (New York, Columbia
    University Press, 1981), p. 72.
Buirette, Patricia, "Genèse d'un droit fluvial international général" (1991) 95
    Rev. Gen. Dr. Int'l Pub. 5.
Bulloch, John and Adel Darwish, *Water Wars: Coming Conflicts in the Middle East*
    (London, Victor Gollancz, 1993).
Bundy, Rodman R., "Natural Resource Development (Oil and Gas) and Boundary
    Disputes" in Gerald H. Blake *et al.* (eds.), *The Peaceful Management of
    Transboundary Resources* (London, Graham & Trotman, 1995), p. 23.

Burness, H. Stuart and James P. Quirk, "Water Law, Water Transfers, and Economic Efficiency: The Colorado River" (1980) 23 J. Law & Econ. 111.

Burton, Lloyd and Chris Cocklin, "Water Resource Management and Environmental Policy Reform in New Zealand: Regionalism, Allocation, and Indigenous Relations" (1996) 7 Colo. J. Int'l Envt'l L. & Pol'y 75.

Byers, Michael, *Custom, Power, and the Power of Rules* (Cambridge, Cambridge University Press, 1999).

Caflisch, Lucius, "Regulation of the Uses of International Watercourses" in Salman M. A. Salman and Laurence Boisson de Chazournes (eds.), *International Watercourses: Enhancing Cooperation and Managing Conflict* (Washington, DC, World Bank, 1998), p. 3.

"Sic utere tuo ut alienum non laedas: Règle prioritaire ou élément servant à mesurer le droit de participation équitable et raisonnable à l'utilisation d'un cours d'eau international" in Alexander von Ziegler and Thomas Burckhardt (eds.), *Internationales Recht auf See und Binnengewässern* (1993), p. 27.

"Règles générales du droit des cours d'eau internationaux" (1989-VII) 219 *Recueil des Cours* 9.

Calabresi, Guido and A. Douglas Melamed, "Property Rules, Liability Rules and Inalienability: One View of the Cathedral" (1972) 85 Harv. L. Rev. 1089.

Calvert, Randy, Mathew D. McCubbins and Barry R. Weingast, "A Theory of Political Control and Agency Discretion" (1989) 33 Am. Jur. Pol. Sci. 588.

Cameron, James and Ruth Mackenzie, "Access to Environmental Justice and Procedural Rights in International Institutions," in Alan E. Boyle and Michael R. Anderson (eds.), *Human Rights Approaches to Environmental Protection* (Oxford, Clarendon Press, 1996), p. 129.

Canter, Larry, Konrad Ott, and Donald A. Brown, "Protection of Marine and Freshwater Resources" in John Lemons and Donald A. Brown (eds.), *Sustainable Development: Science, Ethics, and Public Policy* (Dordrecht, Kluwer Academic Publishers, 1995).

Caponera, Dante A., *Principles of Water Law and Administration: National and International* (Rotterdam and Brookfield, VT, A. A. Balkema, 1992).

Caponera, Dante and Dominique Alheritiere, "Principles for International Groundwater Law" (1978) 18 Nat. Res. J. 589.

Capotorti, Francesco, *Study on the Rights of Persons Belonging to Ethnic, Religious and Linguistic Minorities* (New York, United Nations, 1979) E/CN.4/Sub.2/384/Rev.1.

Caron, David, "The Frog that Wouldn't Leap: The International Law Commission and its Work on International Watercourses" (1992) 3 Colo. J. Int'l Envt'l L. & Pol'y 269.

Carter, Francis W. and David Turnock (eds.), *Environmental Problems in Eastern Europe* (London, Routledge, 1993).

de Castro, Paulo Canelas, "The Judgment in the Case Concerning the
    Gabcikovo–Nagymaros Project: Positive Signs for the Evolution of
    International Water Law" (1997) 8 Yb. Int'l Envt'l L. 21.
Charney, Jonathan I., "Universal International Law" (1993) 87 AJIL 529.
Chayes, Abram and Antonia Handler Chayes, *The New Sovereignty: Compliance
    with International Regulatory Agreements* (Cambridge, MA, Harvard University
    Press, 1995).
Chinkin, Christine, "Human Rights and the Politics of Representation: Is there
    a Role for International Law?" in Michael Byers (ed.), *The Role of Law in
    International Politics* (Oxford, Oxford University Press, 2000), p. 131.
    "Enhancing the International Law Commission's Relationships with Other
    Law-Making Bodies and Relevant Academic and Professional Institutions",
    in *Making Better International Law: The International Law Commission* (United
    Nations, 1998), p. 50.
Churchill R. R. and A. V. Lowe, *The Law of the Sea* (Manchester, UK, Manchester
    University Press, 1988).
Clagett, Brice M., "Survey of Agreements Providing for Third-Party Resolution
    of International Water Disputes" (1961) 55 AJIL 645.
Clark, Colin W., "Restricted Access to Common-Property Fishery Resources: A
    Game-Theoretic Analysis" in Pan-Tai Liu (ed.), *Dynamic Optimization and
    Mathematical Economics* (New York, Plenum Press, 1980), p. 117.
Coase, Ronald H., "The Problem of Social Cost" (1960) 3 J. Law & Econ. 1.
Collins, Robert O., *The Waters of the Nile: Hydropolitics and the Jonglei Canal,
    1900–1988* (Oxford, Clarendon Press, 1990).
Conforti, Benedetto, *International Law and the Role of Domestic Legal Systems*
    (Boston, M. Nijhoff, 1993).
Conybeare, John A. C., "International Organization and the Theory of Property
    Rights" (1980) 34 Int'l Org. 307.
Cooter, Robert D., "Decentralized Law for a Complex Economy: The Structural
    Approach to Adjudicating the New Law Merchant" (1996) 144 U. Pa. L. Rev.
    1643.
Cooter, Robert D. and Lewis Kornhauser, "Can Litigation Improve the Law
    Without the Help of Judges?" (1980) 9 J. Leg. Stud. 139.
Cotterrell, Roger, *The Sociology of Law* (2nd edn, London, Butterworths, 1992).
Cover, Robert M., "The Origins of Judicial Activism in the Protection of
    Minorities" (1982) 91 Yale L.J. 1287.
Craig Paul P. and Grainne de Burca, *EU Law* (2nd edn, New York, Oxford
    University Press, 1998).
Cross, Frank B., "Paradoxical Perils of the Precautionary Principle" (1996) 53
    Wash. & Lee L. Rev. 851.
    "When Environmental Regulations Kill" (1995) 22 Ecology L.Q. 729.
de la Cruz, Hector Bartolomei *et al.*, *The International Labor Organization: The
    International Standards System and Basic Human Rights* (Boulder, CO, Westview
    Press, 1996).

Davis, Morton D., *Game Theory: A Nontechnical Introduction* (Revised edn, New York, Basic Books, 1983).

Deckmyn, Veerle and Ian Thomson (eds.), *Openness and Transparency in the European Union* (Maastricht, European Institute of Public Administration, 1998).

Dellapenna, Joseph W. (ed.), *The Regulated Riparian Model Water Code: Final Report of the Water Laws Committee of the Water Resources Planning and Management Division of the American Society of Civil Engineers* (New York, American Society of Civil Engineers, 1997).

Dodge, Toby and Tariq Tell, "Peace and the Politics of Water in Jordan" in J. A. Allan (ed.), *Water, Peace, and the Middle East* (London, Tauris Academic Studies, 1996), p. 169.

Doherty, Kathryn B., "Jordan Waters Conflict" in 1965 *International Conciliation*, No. 553 (New York, Carnegie Endowment for International Peace, 1965).

Dorf, Michael C. and Charles F. Sabel, "A Constitution of Democratic Experimentalism" (1998) 98 Colum. L. Rev. 267.

Douglas-Scott, Sionaidh, "Environmental Rights in the European Union – Participatory Democracy or Democratic Deficit?" in Alan E. Boyle and Michael R. Anderson (eds.), *Human Rights Approaches to Environmental Protection* (Oxford, Clarendon Press, 1996), p. 109.

Downs, George W. and David M. Rocke, *Optimal Imperfection?* (Princeton, NJ, Princeton University Press, 1995).

Downs, George W. *et al.*, "The Transformational Model of International Regime Design: Triumph of Hope or Experience?" (2000) 30 Colum. J. Trans. L. 465.

Driver, G. R. and John C. Miles (eds.), *The Babylonian Laws: Ancient Codes and Laws of the Near East* (2 vols., Oxford, Clarendon Press, 1952–5).

Duane, Timothy P., "Community Participation in Ecosystem Management" (1997) 24 Ecology L.Q. 771.

Dudley, Norman J., "Water Allocation by Markets, Common Property and Capacity Sharing: Companions or Competitors?" (1992) 32 Nat. Res. J. 757.

Dufour, Sophie, "The Legal Impact of the Canada–United States Free Trade Agreement on Canadian Water Exports" (1993) 34(2) *Les cahiers de droit* 705.

Duncan, John C., "Multicultural Participation in the Public Hearing Process: Some Theoretical, Pragmatical and Analytical Considerations" (1999) Colum. J. Envt'l L. 169.

van Dunne, Jan M., "Liability in Tort for the Detrimental Use of Fresh Water Resources under Dutch Law in Domestic and International Cases" in Edward H. P. Brans *et al.* (eds.), *The Scarcity of Water* (London, Kluwer Law International, 1997), p. 196.

Dunoff, Jeffrey L., "From Green to Global: Toward the Transformation of International Environmental Law" (1995) 19 Harv. Envtl. L. Rev. 241.

Dunoff, Jeffrey L. and Joel P. Trachtman, "Economic Analysis of International Law" (1999) 24 Yale J. Int'l L. 1.

Dupuy, Pierre-Marie, "La coopération régionale transfrontalière et le droit international" (1997) 23 Ann. Franc. Dr. Int'l 837.

Eaton, David J., and David Hurlbut, *Challenges in the Binational Management of Water Resources in the Rio Grande/Rio Bravo* (Austin, TX, Lyndon B. Johnson School of Public Affairs, University of Texas at Austin, 1992).

Economic Commission of Europe, *Two Decades of Co-operation on Water* (1988).

Ellickson, Robert C., *Order without Law: How Neighbors Settle Disputes* (Cambridge, MA, Harvard University Press, 1991).

"A Hypothesis of Wealth-Maximizing Norms: Evidence from the Whaling Industry," (1989) 5 J. Law, Econ. & Org. 83.

Elliott, Lorraine M., *International Environmental Politics: Protecting the Antarctic* (New York, St. Martin's Press, 1994).

Elmusa, Sharif S., "Dividing the Common Palestinian–Israeli Waters: An International Water Law Approach" (Spring 1993) 22 J. Palestine Studies 3.

Elster, Jon, *Deliberative Democracy* (New York, Cambridge University Press, 1998).

Ely, John Hart, *Democracy and Distrust* (Cambridge, MA, Harvard University Press, 1980).

Epstein, Richard A., "Exit Rights under Federalism" (1992) 55 *Law and Contemporary Problems* 147.

Esty, Daniel C., "Non-Governmental Organizations at the World Trade Organization: Cooperation, Competition, or Exclusion" (1998) 1 J. Int'l Econ. L. 123.

"Linkages and Governance: NGOs at the World Trade Organization" (1998) 19 U. Pa. J. Int'l Econ. L. 709.

"Revitalizing Environmental Federalism" (1996) 95 Mich. L. Rev. 570.

Evans, Malcolm D., *Relevant Circumstances and Maritime Delimitation* (Oxford, Clarendon Press, 1989).

Evans, Peter B., "Building an Integrative Approach to International and Domestic Politics in Double-Edged Diplomacy" in Peter B. Evans, Harold K. Jacobson and Robert D. Putnam (eds.), *Double-Edged Diplomacy* (Berkeley, University of California Press, 1993), p. 397.

Ezrahi, Yaron, *The Descent of Icarus* (Cambridge, MA, Harvard University Press, 1990).

Falk, Richard A., *The Role of Domestic Courts in the International Legal Order* (Syracuse, NY, Syracuse University Press, 1964).

Farber, Daniel A. and Paul A. Hemmersbaugh, "The Shadow of the Future: Discount Rates, Later Generations, and the Environment" (1993) 46 Vand. L. Rev. 267.

Fauré, Guy Olivier and Jeffrey Z. Rubin (eds.), *Culture and Negotiation: The Resolution of Water Disputes* (Newbury Park, CA, Sage Publications, 1993).

Fearon, James D., "Deliberation as Discussion", in Jon Elster (ed.), *Deliberative Democracy* (Cambridge, Cambridge University Press, 1998), p. 44.

Feitelson, Eran and Qasem Hassan Abdul-Jaber, *Prospects for Israeli–Palestinian Cooperation in Wastewater Treatment and Re-use in the Jerusalem Region* (Jerusalem, Jerusalem Institute for Israel Studies, Palestinian Hydrology Group, 1997).

Fisher, Roger and William Ury, *Getting to Yes: Negotiating Agreement without Giving In* (Boston, Houghton Mifflin, 1981).

Fiss, Owen M., "The Supreme Court, 1978 Term – Forward: The Forms of Justice" (1979) 93 Harv. L. Rev. 1.

Fisseha, Yimer, "State Succession and the Legal Status of International Rivers" in Ralph Zacklin and Lucius Caflisch (eds.), *The Legal Regime of International Rivers and Lakes* (The Hague, M. Nijhoff, 1981), p. 177.

Fitzmaurice, Malgosia, "The Finnish–Swedish Frontier Rivers Commission" (1992) 5 Hague YB Int'l L. 33.

Fox, Hazel, *Joint Development of Offshore Oil and Gas* (2 vols., London, British Institute of International and Comparative Law, 1989).

Francis, George, "Ecosystem Management" (1993) 33 Nat. Res. J. 316.

Franck, Thomas M., *Fairness in International Law and Institutions* (Oxford, Clarendon Press, 1995).

The Power of Legitimacy among Nations (New York, Oxford University Press, 1990).

"The Courts, the State Department, and National Policy: A Criterion for Judicial Abdication" (1960) 44 Minn. L. Rev. 1101.

Freestone, David and Ellen Hey, "Origins and Development of the Precautionary Principle" in David Freestone and Ellen Hey (eds.), *The Precautionary Principle and International Law: The Challenge of Implementation* (The Hague, Kluwer Law International, 1996), p. 3.

Frey, Bruno S., "The Public Choice of International Organization in Perspectives" in Dennis C. Mueller (ed.), *Perspectives on Public Choice* (New York, Cambridge University Press, 1995), p. 106.

Frischwasser-Ra'anan, H. F., *The Frontiers of a Nation: A Re-examination of the Forces which Created the Palestine Mandate and Determined its Territorial Shape* (London, Batchworth Press, 1955).

Frowein, Jochen Abr., "The Protection of Property" in R. St. J. Macdonald *et al.* (eds.), *The European System for the Protection of Human Rights* (Dordrecht, M. Nijhoff, 1993), p. 515.

"Germany," in Francis G. Jacobs and Shelley Roberts (eds.), *The Effect of Treaties in Domestic Law* (London, Sweet & Maxwell, 1987), p. 63.

Frowein, Jochen Abr. and Michael J. Hahn, "The Participation of Parliament in the Treaty Process in the Federal Republic of Germany" (1991) 67 Chi. Kent L. Rev. 361.

Fuentes, Ximena, "The Criteria for the Equitable Utilization of International Rivers" (1996) 67 Brit. Yb. Int'l L. 337.

Gaja, Giorgio, "Italy" in Francis G. Jacobs and Shelley Roberts (eds.), *The Effect of Treaties in Domestic Law* (London, Sweet & Maxwell, 1987), p. 87.

Galambos, Judit, "Political Aspects of an Environmental Conflict: The Case of the Gabcikovo–Nagymaros Dam System" in Jurki Kakonen (ed.), *Perspectives on Environmental Conflict and International Politics* (London, Pinter Publishers, 1992), p. 72.

Garfinkle, Adam, *War, Water, and Negotiation in The Middle East: The Case of the Palestine–Syria Border, 1918–1923* (Tel Aviv University, Moshe Dayan Center for Middle Eastern and African Studies, 1994).

Garrett, Geoffrey, "Capital Mobility, Trade, and the Domestic Politics of Economic Policy," in Robert O. Keohane and Helen V. Milner (eds.), *Internationalization and Domestic Politics* (New York, Cambridge University Press, 1996), p. 79.

Garrett, Geoffrey and Peter Lange, "Internationalization, Institutions and Political Change" in Robert O. Keohane and Helen V. Milner (eds.), *Internationalization and Domestic Politics* (New York, Cambridge University Press, 1996), p. 48.

Gaubatz, Kurt T., "Democratic States and Commitment in International Relations" (1996) 50 Int'l Org. 108.

Geertz, Clifford, "Organization of the Balinese Subak" in E. Walter Coward, Jr. (ed.), *Irrigation and Agricultural Development in Asia: Perspectives from the Social Sciences* (Ithaca, NY, Cornell University Press, 1980).

Geisse, Monica Gangas and Herman Santis Arenas, "Chile–Bolivia Relations: The Lauca River Water Resources', in Gerald H. Blake *et al.* (eds.), *The Peaceful Management of Transboundary Resources* (London, Graham & Trotman, 1995), p. 277.

Gellner, Ernest, *Nations and Nationalism* (London, Cornell University Press, 1983).

Gillespie, Alexander, *International Environmental Law Policy and Ethics* (Oxford, Clarendon Press, 1997).

Gleick, Peter H., "An Introduction to Global Fresh Water Issues" in Peter H. Gleick (ed.), *Water in Crisis: A Guide to the World's Fresh Water Resources* (New York, Oxford University Press, 1993).

Godana, Bonaya Adhi, *Africa's Shared Water Resources* (London, F. Pinter, 1985).

Goetz, Charles J. and Robert E. Scott, "Principles of Relational Contracts" (1981) 67 Va. L. Rev. 1089.

Goldberg, David, "World Bank Policy on Projects on International Waterways in the Context of Emerging International Law and the Work of the International Law Commission" in Gerald H. Blake *et al.* (eds.), *The Peaceful Management of Transboundary Resources* (London, Graham & Trotman, 1995), p. 153.

Goldie, Louis F. E., "Reconciling Values of Distributive Equity and Management Efficiency in the International Commons' in René-Jean Dupuy (ed.), *The Settlement of Disputes on the New Natural Resources* (The Hague, M. Nijhoff, 1983), p. 335.

Goldsmith, Edward, and Nicholas Hildyard, *The Social and Environmental Effects of Large Dams* (San Francisco, Sierra Club Books, 1984).

Goodwin-Gill, Guy S., "Obligations of Conduct and Result" in Philip Alston and Katarina Tomasevski (eds.), *The Right to Food* (Boston, M. Nijhoff, 1984), p. 111.

Gordon, H. Scott, "The Economic Theory of a Common-Property Resource: The Fishery" (1954) 62 J. Pol. Econ. 124.

Goswami, Subir, *Politics in Law Making* (New Delhi, Ashish Publishing House, 1986).

Grieco, Joseph M., *Cooperation among Nations: Europe, America, and Non-Tariff Barriers to Trade* (Ithaca, NY, Cornell University Press, 1990).

Griffin, Ronald C. and Fred O. Boadu, "Water Marketing in Texas: Opportunities for Reform" (1992) 32 Nat. Res. J. 265.

Griffin, William L., "The Use of Waters of International Drainage Basins Under Customary International Law" (1959) 53 AJIL 50.

Grotius, Hugo, *The Freedom of The Seas*, trans. by Ralph von Deman Magoffin, James Brown Scott (ed.) (Oxford University Press, 1916).
  De ivre belli ac pacis libri tres. In quibus jus naturae & gentium: item juris publici praecipua explicantur (1632).

Guendling, Lothar, "Our Responsibility to Future Generations" (1990) 84 AJIL 207.

Gulhati, N. D., *Development of Inter-State Rivers: Law and Practice of India* (Bombay, Allied Publ. 1972).

Gulmann, Claus, "Denmark" in Francis G. Jacobs and Shelley Roberts (eds.), *The Effect of Treaties in Domestic Law* (London, Sweet & Maxwell, 1987), p. 29.

Guzman, Andrew T., "Is International Antitrust Possible?" (1998) 73 NYU L. Rev. 1501.

Haas, Peter M., "Introduction: Epistemic Communities and International Policy Coordination" (1992) 46 Int'l Org. 1.
  (ed.) "Knowledge, Power and International Policy Coordination" (1992) 46 Int'l Org. 1 (Special Issue).

Haas, Peter M., Robert O. Keohane, and Marc A. Levy (eds), *Institutions for the Earth,* (Cambridge, MA, MIT Press, 1993).

Habermas, Juergen, *The Theory of Communicative Action* (Boston, Beacon Press, vol. I, 1984; vol. II, 1987).

HaCohen, Mordechai, *Halachot veHalichot (Norms and Manners)* (Jerusalem, 1975, in Hebrew).

Hafner, Gerhard, "The Optimum Utilization Principle and the Non-Navigational Uses of Drainage Basins" (1993) 45 Austrian J. Publ. Int'l Law 113.

Haggard, Stephan and Sylvia Maxfield, "The Political Economy of Financial Internationalization in the Developing World" in Robert O. Keohane and Helen V. Milner (eds.), *Internationalization and Domestic Politics* (New York, Cambridge University Press, 1996), p. 209.

Haggard, Stephan and Beth A. Simmons, "Theories of International Regimes" (1987) 41 Int'l Org. 491.

Handl, Gunther, "The International Law Commission's Draft Articles on the Law of International Watercourses (General Principles and Planned Measures): Progressive or Retrogressive Development of International Law?" (1992) 3 Colo. J. Int'l Envt'l L. & Pol'y 123.

"The Principle of 'Equitable Use' as Applied to Internationally Shared Natural Resources: Its Role in Resolving Potential International Disputes over Transfrontier Pollution" (1978) 14 Rev. Belge Dr. Int'l 40.

"Balancing of Interests and International Liability for the Pollution of International Watercourses: Customary Principles of Law Revisited" (1975) Can. Yb. Int'l L. 156.

Hanson, Jon D. and Douglas A. Kysar, "Taking Behavioralism Seriously: Some Evidence of Market Manipulation" (1999) 112 Harv. L. Rev. 1422.

Hardin, Garret, "The Tragedy of the Commons" (1968) 162 *Science* 1243.

Hardin, Russell, *Collective Action* (Baltimore, MD, Johns Hopkins University Press, 1982).

Hayton, Robert D., "Observations on the ILC's Draft Rules: Articles 1–4" (1992) 3 Colo. J. Int'l Envt'l L. & Pol'y 31.

Hayton, Robert D. and Albert E. Utton, "Transboundary Groundwater: The Bellagio Draft Treaty" (1989) 29 Nat. Res. J. 663.

Hayward, M., "International Law and the Interpretation of the Canadian Charter of Rights and Freedoms: Uses and Justifications" (1985) 23 U. West. Ontario L. Rev. 9.

Held, David, *Democracy and the Global Order* (Stanford, CA, Stanford University Press, 1995).

Henkin, Louis, *Foreign Affairs and the United States Constitution* (2nd edn, Oxford, Oxford University Press, 1996).

Constitutionalism, Democracy, and Foreign Affairs (New York, Columbia University Press, 1990).

"International Law as Law in the United States" (1984) 82 Mich. L. Rev. 1555.

How Nations Behave (2nd edn, New York, Columbia University Press, 1979).

Hey, Ellen, "Sustainable Use of Shared Water Resources: The Need for a Paradigmatic Shift in International Watercourses Law" in Gerald H. Blake *et al.* (eds.), *The Peaceful Management of Transboundary Resources* (London, Graham & Trotman, 1995), p. 127.

Higgins, Rosalyn, "United Kingdom" in Francis G. Jacobs and Shelley Roberts (eds.), *The Effect of Treaties in Domestic Law* (London, Sweet & Maxwell, 1987) p. 123.

Hilf, Meinhard, "New Frontiers in International Trade: The Role of National Courts in International Trade Relations" (1997) 18 Mich. J. Int'l L. 321.

Hirsch, Moshe, *The Responsibility of International Organizations toward Third Parties: Some Basic Principles* (Dordrecht, M. Nijhoff, 1995).

Hirsch, Philip, "Natural Resource Conflict and 'National Interest' in Mekong Hydropower Development" (1999) 29 Golden Gate U.L. Rev. 399.

Hirschman, Albert O., *Exit, Voice and Loyalty* (Cambridge, MA, Harvard University Press, 1970).

Hobbes, Thomas, *Leviathan* (W.G. Pogson Smith ed., Oxford, Clarendon Press, 1909).

Hobsbawm, Eric J., *Nations and Nationalism Since 1780* (2nd ed., New York, Cambridge University Press, 1992).

van Hoof, Godfried, "The Legal Nature of Economic, Social and Cultural Rights: A Rebuttal of Some Traditional Views" in Philip Alston and Katarina Tomasevski (eds.), *The Right to Food* (Boston, M. Nijhoff, 1984) p. 97.

Horn, Henrik, Petros C. Mavroidis and Hakan Nordstroem, "Is the Use of the WTO Dispute Settlement System Biased?" CEPR Discussion Paper No. 2859 (London, 2001).

Hornstein, Donald T., "The Political Origins of Modern Environmental Law: Self-interest, Politics, and the Environment – A Response to Professor Schroeder" (1998) 9 Duke Env. L & Pol'y F 61.
  "Reclaiming Environmental Law: A Normative Critique of Comparative Risk Analysis" (1992) 92 Colum. L. Rev. 562.

Horowitz, Donald L., *Ethnic Groups in Conflict* (Berkeley, University of California Press, 1985).

Horwitz, Morton J., *The Transformation of American Law 1780–1860* (Cambridge, MA, Harvard University Press, 1977).

Howell, Paul, Michael Lock and Stephen Cobb (eds.), *The Jonglei Canal: Impact and Opportunity* (Cambridge, Cambridge University Press, 1988).

Howse, Robert, "Democracy, Science and Free Trade: Risk Regulation on Trial at the World Trade Organization" (unpublished manuscript on file with author, 1999).

Huffman, James L., "Institutional Constraints on Transboundary Water Marketing" in Terry L. Anderson and Peter J. Hill (eds.), *Water Marketing – The Next Generation* (Lanham, MD, Rowman & Littlefield, 1997), p. 31.

Hunt, Constance D., "Implementation: Joint Institutional Management and Remedies in Domestic Tribunals (Articles 26–28 and 30–32)" (1992) 3 Colo. J. Int'l Envt'l L. & Pol'y 281.

Hurrell, Andrew and Benedict Kingsbury, "Introduction", in Hurrell and Kingsbury (eds.) *The International Politics of the Environment: Actors, Interests, and Institutions* (Oxford, Clarendon Press, 1992).

Hurst, Phillip, *Rainforest Politics: Ecological Destruction in South-East Asia* (London and Atlantic Highlands, NJ, Zed Books, 1990), p. 197.

Iglesias, Maria Teresa Ponte, "Les accords conclus par les autorités locales de différents États sur l'utilisation des eaux frontalières dans le cadre de la coopération transfrontalière" in (2/1995) Schweiz. Z. Int'l. Europ. R. 103.

International Federation of Red Cross and Red Crescent Societies, *World Disasters Report* (Dordrecht, Martinus Nijhoff, 1999).

Israel, Morris and Jay R. Lund, "Recent California Water Transfers: Implications for Water Management" (1995) 35 Nat. Res. J. 1.

Israel State Comptroller, Report on Mekorot Water Company Ltd., (1995, in Hebrew).

Report on the Management of the Water Economy in Israel (1990, in Hebrew).

Iwasawa, Yuji, *International Law, Human Rights, and Japanese Law* (Oxford, Oxford University Press, 1998).

Jackson, John H., "United States", in Francis G. Jacobs and Shelley Roberts (eds.), *The Effect of Treaties in Domestic Law* (London, Sweet & Maxwell, 1987), p. 141.

Jain, S. N., Alice Jacob and Subhash C. Jain, *Interstate Water Disputes in India: Suggestions for Reform in Law* (Bombay, NM Tripathi, 1971).

Jennings, Robert Y., "The Identification of International Law," in Bin Cheng (ed.), *International Law: Teaching and Practice* (London, Stevens, 1982), p. 3.

"The Progressive Development of International Law and its Codification" (1947) 24 Brit. Yb Int'l L. 301.

Johnson, Ronald N., Micha Gisser, and Michael Werner, "The Definition of Surface Water Right and Transferability" (1981) 24 J. Law & Econ., 273.

Jolls, Christine, Cass R. Sunstein and Richard Thaler, "A Behavioral Approach to Law and Economics" (1998) 50 Stan. L. Rev. 1471.

Kahn, Peter L., "The Politics of Unregulation: Public Choice and Limits on Government," (1990) 75 Cornell L. Rev. 280.

Kassas, M., "Environmental Aspects of Water Resources Development" in Asit K. Biswas (ed.), *Water Management for Arid Lands in Developing Countries* (Oxford, Pergamon Press, 1980) p. 67.

Kearney, Richard D., "International Watercourses" in René-Jean Dupuy (ed.), *A Handbook on International Organizations* (Dordrecht, M. Nijhoff, 1988), p. 509.

Keohane, Robert O., *International Institutions and State Power* (Boulder, CO, Westview Press, 1989).

After Hegemony (Princeton, NJ, Princeton University Press, 1984).

Keohane, Robert O. and Lisa L. Martin, "The Promise of Institutionalist Theory" (1995) 20 Int'l Sec. 39.

Khouri, Rami G., *The Jordan Valley: Life and Society Below Sea Level* (London, Longmans, 1981).

Kingsbury, Benedict, "Operational Policies of International Institutions as Part of the Law-Making Process: The World Bank and Indigenous Peoples" in Guy S. Goodwin-Gill and Stefan Talmon (eds.), *The Reality of International Law: Essays in Honour of Ian Brownlie* (Oxford, Clarendon Press, 1999), p. 323.

"Claims by Non-State Groups in International Law" (1992) 25 Cornell Int'l L.J. 481.

Kiss, Alexandre C., "Commentaire" (1986) 2–3 Rev. Jur. Envn't 307.

"Commentaire" (1983) 4 Rev. Jur. Envn't 353.

Klemencic, Mladen, "The Effects of War on Water and Energy Resources in Croatia and Bosnia" in Gerald H. Blake *et al.* (eds.), *The Peaceful Management of Transboundary Resources* (London, Graham & Trotman, 1995).

Koh, Harold Hongju, "Commentary: Is International Law Really State Law?" (1998) 111 Harv. L. Rev. 1824.

   "The 'Haiti Paradigm' in United States Human Rights Policy" (1994) 103 Yale L.J. 2391.

   "The Fast Track and United States Trade Policy" (1992) 18 Brook. J. Int'l L. 143.

   "Transnational Public Law Litigation" (1991) 100 Yale L.J. 2347.

Kolars, John, "The Course of Water in the Arab Middle East" (1990) 33 Am. Ar. Aff. 57.

Koremanos, Barbara, Charles Lipson, and Duncan Snidal, *Rational International Institutions* (Rational International Institutions Project, at http://www.harisschool.uchi, 1998).

Koskenniemi, Martti, *From Apology to Utopia* (Helsinki, Finnish Lawyers' Pub. Co., 1989).

Krasner, Stephan D. "Global Communications and National Power, Life on the Pareto Frontier" (1991) 43 *World Politics* 336.

   (ed.), *International Regimes* (Ithaca, NY, Cornell University Press, 1983).

Kremenyuk, Victor A. (ed.), *International Negotiations: Analysis, Approaches, Issues* (San Francisco, Jossey-Bass Publishers, 1991).

Kreps, David M. *et al.*, "Rational Cooperation in the Finitely Repeated Prisoners' Dilemma" (1982) 27 J. Econ. Theo. 245.

Krishna, Raj, "The Evolution and Context of the Bank Policy for Projects on International Waterways" in Salman M. A. Salman and Laurence Boisson de Chazournes (eds.), *International Watercourses: Enhancing Cooperation and Managing Conflict* (Washington, DC, World Bank, 1998), p. 31.

   "International Watercourses: World Bank Experience and Policy" in J. A. Allan and Chibli Mallat (eds.), *Water in the Middle East: Legal, Political and Commercial Implications* (London, I.B. Tauris Publishers, 1995), p. 29.

Kushner, David, "Conflict and Accommodation in Turkish–Syrian Relations" in Moshe Ma'oz and Avner Yaniv (eds.), *Syria under Assad: Domestic Constraints and Regional Risks* (London, Croom Helm, 1986), p. 85.

Lagoni, Rainer, "Oil and Gas Deposits across National Frontiers" (1979) 73 AJIL 215.

Lambton, A. K. S., "Qanat," 4 *Encyclopedia of Islam*, 529–31.

Lammers, J. G., *Pollution of International Watercourses* (The Hague, M. Nijhoff, 1984).

Landes, William M. and Richard A. Posner, "The Independent Judiciary in an Interest-Group Perspective" (1975) 18 J. Law & Econ. 875.

Lapidoth, Ruth, "Equity in International Law" (1987) 22 Israel L. Rev. 161.

Lauterpacht, Elihu, *Aspects of the Administration of International Justice* (Cambridge, Grotius, 1991).

"River Boundaries: Legal Aspects of the Shatt-Al-Arab Frontier" (1960) 9 ICLQ 216.

Lax, David A. and James K. Sebenius, *The Manager as Negotiator* (Washington DC, National Institute for Dispute Resolution, 1985).

LeMarquand, David G., "Preconditions to Cooperation in Canada–United States Boundary Waters" (1986) 26 Nat. Res. J. 221.

International Rivers: The Politics of Cooperation (Vancouver, Westwater Research Centre, University of British Colombia, 1977).

Lemons, John and Donald A. Brown (eds.), *Sustainable Development: Science, Ethics, and Public Policy* (Dordrecht, Kluwer Academic Publishers, 1995).

Levy, Marc A. *et al.*, "Improving the Effectiveness of International Environmental Institutions" in Peter M. Haas, Robert O. Keohane, and Marc A. Levy (eds.), *Institutions for the Earth: Sources of Effective International Environmental Protection*, Global Environmental Accords Series (Cambridge, MA, MIT Press, 1993), p. 397.

Lewinsohn-Zamir, Daphna, "Consumer Preferences, Citizen Preferences, and the Provision of Public Goods" (1998) 108 Yale L.J. 377.

Libecap, Gary D., *Contracting for Property Rights* (Cambridge, Cambridge University Press, 1989).

Lipper, Jerome, "Equitable Utilization" in Albert H. Garretson, Robert D. Hayton, and Cecil J. Olmstead (eds.), *The Law of International Drainage Basins* (Dobbs Ferry, NY, Oceana Publications, 1967), p. 15.

Lohmann, Susanne, "An Information Rationale for the Power of Special Interests" (1998) 92(4) Am. Pol. Sci. Rev. 809.

Maass, Arthur and Raymond L. Anderson, *. . . And The Desert Shall Rejoice: Conflict, Growth and Justice in Arid Environments* (Cambridge, MA, MIT Press, 1978).

MacNeil, Ian R., "Economic Analysis of Contractual Relations: Its Shortfalls and the Need for a 'Rich Classificatory Apparatus'" (1981) 75 Nw U.L. Rev. 1018.

The New Social Contract: An Inquiry into Modern Contractual Relations (New Haven, Yale University Press, 1980).

"The Many Futures of Contract" (1974) 47 S. Cal. L. Rev. 691.

Maktari, A. M. A., *Water Rights and Irrigation Practices in Lahj* (Cambridge, Cambridge University Press, 1971).

Mallat, Chibli, "The Quest for Water Use Principles: Reflections on the *Shari'a* and Custom in the Middle East" in J. A. Allen and Chibli Mallat (eds.), *Water in the Middle East: Legal, Political, and Commercial Implications* (London, I. B. Tauris Publishers, 1995), p. 127.

"Law and the Nile River: Emerging International Rules and the Shari'a" in P. P. Howell and J. A. Allen (eds.), *The Nile: Sharing a Scarce Resource* (Cambridge, Cambridge University Press, 1994), p. 365.

Margolis, Howard, *Dealing with Risk* (Chicago, IL, University of Chicago Press, 1996).

Marshall, Eliot, "A Is for Apple, Alar and . . . Alarmist?" (4 Oct. 1991) 254
    Science 20.

Martin, Lisa L., and Beth A. Simmons, "Theories and Empirical Studies of
    International Institutions" (1998) 52 Int'l Org., 729.

McCaffrey, Stephen S., "A Human Right to Water: Domestic and International
    Implications" (1992) 5 Georgetown Int'l Envt'l L. Rev. 1.

    Special Rapporteur, *Seventh Report on the Law of the Non-Navigational Uses of
    International Watercourses*, UN Doc. A/CN.4/436 (1991).

    Sixth Report on the Law of the Non-Navigational Uses of International Watercourses,
    UN Doc. A/CN.4/427/Add.1 (1990).

    "Second Report on the Law of the Non-Navigational Uses of International
    Watercourses" Doc. A/CN.4/399 (1986) 2 *ILC Yearbook* (Part 1) 87.

    "Legal Problems Relating to the Utilization and Use of International Rivers"
    Doc. A/5409 (1974) 2 *ILC Yearbook* (Part 2) 33.

McCully, Patrick, *Silenced Rivers: The Ecology and Politics of Large Dams* (London and
    Atlantic Highlands, NJ, USA, Zed Books, 1996).

McGinnis, John O., "The Decline of the Western Nation State and the Rise of
    the Regime of International Federalism" (1996) 18 Cardozo L. Rev. 903.

Meyers, Charles J. and Richard A. Posner, *Market Transfer of Water Rights: Towards
    an Improved Market in Water Resources* 290 (National Water Commission,
    Legal Study No. 4, NTIS No. NWC-L-71-009, July 1971).

Michel, Aloys A., *The Indus Rivers: A Study of the Effects of Partition* (New Haven,
    Yale University Press, 1967).

Milner, Helen V., "Rationalizing Politics: The Emerging Synthesis of
    International, American, and Comparative Politics" (1998) 52 Int'l Org. 759.

    Interests, Institutions, and Information: Domestic Politics and International Relations,
    (Princeton, NJ, Princeton University Press, 1997).

Miyoshi, Masahiro, *Considerations of Equity in the Settlement of Territorial and
    Boundary Disputes* (Dordrecht, M. Nijhoff, 1993).

Monna, Stevie C., "A Framework for International Cooperation for the
    Management of the Okavango Basin and Delta", Ramsar COP7 DOC. 20.5
    (1999) http://www.ramsar.org/cop7_doc_20.5_e.htm

Morrison, Edward R., "Judicial Review of Discount Rates Used in Regulatory
    Cost-Benefit Analysis" (1998) 65 U. Chi. L. Rev. 1333.

Myers, Norman, "The Anatomy of Environmental Action: The Case of Tropical
    Deforestation" in Andrew Hurrell and Benedict Kingsbury (eds.), *The
    International Politics of the Environment: Actors, Interests, and Institutions* (Oxford,
    Oxford University Press, 1992), p. 430.

Nelson, L. D. M., "The Roles of Equity in the Delimitation of Maritime
    Boundaries" (1990) 84 AJIL 837.

Neuman, Gerald L., "Sense and Nonsense about Customary International Law:
    A Response to Professors Bradley and Goldsmith" (1997) 66 Fordham L.
    Rev. 371.

Nichols, Philip M., "Participation of Nongovernmental Parties in the World Trade Organization: Extension of Standing in World Trade Organization Disputes to Nongovernment Parties" (1996) 17 U. Pa. J. Int'l Econ. L. 295.

Niskanen, William A., *Bureaucracy and Public Economics* (Brookfield, VT, E. Elgar, 1994).

Nollkaemper, André, "'What You Risk Is What You Value', and Other Dilemmas Encountered in the Legal Assaults on Risks" in David Freestone and Ellen Hey (eds.), *The Precautionary Principle and International Law: The Challenge of Implementation* (The Hague, Kluwer Law International, 1996), p. 73.

The Legal Regime for Transboundary Water Pollution: Between Discretion and Constraint (Dordrecht, M. Nijhoff, 1993).

Note, "Judicial Enforcement of International Law against the Federal and State Governments" (1991) 104 Harv. L. Rev. 1269.

O'Connell, Daniel P., *State Succession in Municipal and International Law*, vol. II (Cambridge, Cambridge University Press, 1967).

Olson, Mancur, *The Rise and Decline of Nations* (New Haven, Yale University Press, 1982).

The Logic of Collective Action (Cambridge, MA, Harvard University Press, 1965).

Ong, David M., "Joint Development of Common Offshore Oil and Gas Deposits: 'Mere' State Practice or Customary International Law?" (1999) 93 AJIL 771.

Osborne, Martin J. and Ariel Rubinstein, *A Course in Game Theory* (Cambridge, MA, MIT Press, 1994).

Ostrom, Elinor, *Crafting Institutions for Self-Governing Irrigation Systems* (San Francisco, ICS Press, 1992).

Governing the Commons: The Evolution of Institutions for Collective Action (Cambridge, Cambridge University Press, 1990).

Paine, Robert, *Dam a River, Dam a People? Saami (Lapp) Livelihood and the Alta/Kautokeino Hydro-Electric Project and the Norwegian Parliament*, IWGIA document (Copenhagen, International Work Group for Indigenous Affairs, 1982).

Palmer, Geoffrey, "New Ways to Make International Environmental Law" (1992) 86 AJIL 259.

Pashovski, Slavi, "Minorities in Bulgaria" in John Packer and Kristian Myntti (eds.), *The Protection of Ethnic and Linguistic Minorities in Europe* (Turku, Finland, Institute for Human Rights, Abo Akademi University, 1993), p. 67.

Patai, Refael, *Ha-Maim (The Water)* (Tel-Aviv, 1936, in Hebrew).

Paul, Joel R., "The Geopolitical Constitution: Executive Expediency and Executive Agreements" (1998) 86 Calif. L. Rev. 671.

Pearce, Fred, *The Damned: Rivers, Dams, and the Coming World Water Crisis* (London, Bodley Head, 1992).

Green Warriors (London, Bodley Head, 1991).

Peterson, M. J., "International Fisheries Management" in Peter M. Haas, Robert O. Keohane, and Marc A. Levy (eds.), *Institutions for the Earth* (Cambridge, MA, MIT Press, 1993), p. 249.

Petts, Geoffrey E., *Impounded Rivers: Perspectives for Ecological Management, Environmental Monographs and Symposia* (Chichester, Wiley, 1984).

Ponting, Clive, *A Green History of the World* (London, Sinclair-Stevenson, 1991).

Popovic, Neil A. F., "The Right to Participate in Decisions that Affect the Environment" (1993) 10 Pace Envtl. L. Rev. 683.

Posner, Eric A., "Law, Economics, and Inefficient Norms" (1996) 144 U. Pa. L. Rev. 1697.

Powell, Robert, "Absolute and Relative Gains in International Relations Theory" (1991) 84(4) Am. Pol. Sci. Rev. 1303.

Propp, William Henry, "Water in the Wilderness: The Mythological Background of a Biblical Motif" (Ph.D., Harvard University, 1985).

Putnam, Robert D., "Diplomacy and Domestic Politics: The Logic of Two-Level Games" (1988) 42 Int'l Org. 427.

Putnam, Susan W. and Jonathan Baert Wiener, "Seeking Safe Drinking Water" in John D. Graham and Jonathan Baert Wiener (eds.), *Risk Versus Risk: Tradeoffs in Protecting Health and the Environment* (Cambridge, MA, Harvard University Press, 1995), p. 124.

Raiffa, Howard, *The Art and Science of Negotiation* (Cambridge, MA, Belknap Press of Harvard University Press, 1982).

Ramseyer, J. Mark, "The Puzzling (In)dependence of courts: A Comparative Approach" (1994) 23 J. Leg. Stud. 721.

Raustiala, Kal, "The Domestic Politics of Global Biodiversity Protection in the United Kingdom and the United States" in Miranda A. Schreurs and Elizabeth Economy (eds.), *The Internationalization of Environmental Protection* (Cambridge, Cambridge University Press, 1997), p. 42.

Note, "The 'Participatory Revolution' in International Environmental Law" (1997) 21 Harv. Envtl. L. Rev. 537.

"International 'Enforcement of Enforcement' under the North American Agreement on Environmental Cooperation" (1996) 36 Va. J. Int'l L. 721.

Rawls, John, *A Theory of Justice* (Oxford, Clarendon Press, 1972).

A Theory of Justice (Cambridge, MA, Belknap Press of Harvard University Press, 1971).

Reich, Arie, "From Diplomacy to Law: The Judicization of International Trade Relations" (1997) 17 J. Int'l L. Bus. 775.

Reisman, W. Michael, "An International Farce: The Sad Case of the PLO Mission" (1989) 14 Yale J. Int'l L. 412.

"The Cult of Custom in the Late 20th Century" (1987) 17 Cal. W. Int'l L.J. 133.

Research Center for International Law, University of Cambridge, *International Boundary Cases: The Continental Shelf* (Cambridge, Grotius, 1992).

Revesz, Richard L., "Rehabilitating Interstate Competition: Rethinking the 'Race-to-the-Bottom' Rationale for Federal Environmental Regulation" (1992) 67 NYU L. Rev. 1210.

Richards, John E., "Towards a Positive Theory of International Institutions: Regulating International Aviation Markets" (1999) 53 Int'l Org. 1.

Riesenfeld, Stefan A. and Frederick M. Abbott, "Foreword: Symposium on Parliamentary Participation in the Making and Operation of Treaties" (1991) 67 Chi. Kent L. Rev. 293.

  "The Scope of US Senate Control over the Conclusion and Operation of Treaties" (1991) 67 Chi. Kent L. Rev. 571.

Risse-Kappen, Thomas, "Structures of Governance and Transnational Relations: What We Have Learned?" in Thomas Risse-Kappen (ed.), *Bringing Transnational Relations Back In* (Cambridge, Cambridge University Press, 1995), p. 280.

Rivera, José A., "Irrigation Communities of the Upper Rio Grande Bioregion: Sustainable Use in the Global Context" (1996) 36 Nat. Res. J. 491.

Robinson, Nicholas A. (ed.), *Agenda 21 & the UNCED Proceedings* (6 vols., New York, Oceana Publications, 1992–3).

Robson, Charles, "Transboundary Petroleum Reservoirs: Legal Issues and Solutions" in Gerald H. Blake *et al.* (eds.), *The Peaceful Management of Transboundary Resources* (London, Graham & Trotman, 1995), p. 3.

Rodriguez, Daniel B., "The Role of Legal Innovation in Ecosystem Management: Perspectives from American Local Government Law" (1997) 24 Ecology L.Q. 745.

Ron, Zvi Y. D., "Qantas and Spring Flow Tunnels in the Holy Land" in Peter Beaumont, Michael Bonnie and Keith McLachlan (eds.), *Qantas, Kariz and Khattara: Traditional Water Systems in the Middle East and North Africa* (London, Middle East & North African Studies Press, 1989), p. 211.

  "Development and Management of Irrigation Systems in Mountain Regions of the Holy Land" (1985) 10 Trans. Inst. Br. Geogr. 149.

  "The Utilization of Springs for Irrigated Agriculture in the Judea Mountains" in Avshalom Shmueli, David Grossman, and Rehav'am Ze'evi, (eds.), *Judea and Samaria* (2 vols., Jerusalem, Canaan Publishing House, 1977, in Hebrew) vol. I, p. 230.

  "Battir – The Village and the System of Irrigated Terraces" (1968) 10 *Teva va-Arets* 112 (in Hebrew).

Rose, Carol M., "Rethinking Environmental Controls: Management Strategies for Common Resources" 1991 Duke L.J. 1.

Rose-Ackerman, Susan, *Corruption and Government: Causes, Consequences, and Reform* (Cambridge, Cambridge University Press, 1999).

Rosenbaum, Michael D., "Domestic Bureaucracies and the International Trade Regime: The Law and Economics of Administrative Law and Administratively-Imposed Trade Barriers" (Discussion Paper No. 250 1/99, The Center for Law, Economics and Business, Harvard Law School).

Rosenne, Shabtai, "The International Law Commission, 1949-1959" (1960) 36 Brit. Yb Int'l L. 107.

Rosenstock, Robert, *Second Report on the Law of the Non-Navigational Uses of International Watercourses*, A/CN.4/462 (1994).

Rossi, Jim, "Participation Run Amok: The Costs of Mass Participation for Deliberative Agency Decisionmaking" (1997) NW U.L. Rev. 173.

Rozas, Jose C. F., "La succession d'Etats en matière de conventions fluviales" in Ralph Zacklin and Lucius Caflisch (eds.), *The Legal Regime of International Rivers and Lakes* (The Hague, M. Nijhoff, 1981), p. 127.

Rubin, Paul H., "Why Is the Common Law Efficient?" (1977) 6 J. Leg. Stud. 51.

Sadik, Abdul-Karim and Shawki Barghouti, "The Water Problems of the Arab World: Management of Scarce Resources" in Peter Rogers and Peter Lydon (eds.), *Water in the Arab World* (Cambridge, MA, Cambridge University Press, 1994), p. 1.

Sadurska, Romana and Christine M. Chinkin, "The Collapse of the International Tin Council: A Case of State Responsibility?" (1990) 30 Va. J. Int'l L. 845.

Salman, Salman M. A., "Sharing the Ganges Waters between India and Bangladesh: An Analysis of the 1996 Treaty" in Salman M. A. Salman and Laurence Boisson de Chazournes (eds.), *International Watercourses: Enhancing Cooperation and Managing Conflict, Proceedings of a World Bank Seminar* (Washington, DC, World Bank, 1998).

The Legal Framework for Water Users' Associations: A Comparative Study (Washington, DC, World Bank, 1997).

Samson, Klaus, "The Standard-Setting and Supervisory System of the International Labour Organisation" in Raija Hanski and Markku Suksi (eds.), *An Introduction to the International Protection of Human Rights* (Turku/Abo, Institute for Human Rights, Abo Akademi University, 1997), p. 149.

Sandler, Todd, *Global Challenges: An Approach to Environmental, Political, and Economic Problems* (Cambridge, Cambridge University Press, 1997).

Collective Action: Theory and Applications (Ann Arbor, University of Michigan Press, 1992).

Sands, Phillipe, *Principles of International Environmental Law* (Manchester, NY, Manchester University Press, 1995).

Sax, Joseph L., Robert H. Abrams, and Barton H. Thompson, Jr., *Legal Control of Water Resources: Cases and Materials* (2nd edn, St. Paul, MN, West Publishing Company, 1991).

Schachter, Oscar, *International Law in Theory and Practice* (Boston, M. Nijhoff, 1991).

"The Emergence of International Environmental Law" (1991) 44 J. Int'l Aff. 457.

Scheling, Thomas C., *Arms and Influence* (New Haven, Yale University Press, 1966).

Schermers, Henry G. and Niels M. Blokker, *International Institutional Law: Unity within Diversity* (3rd edn, The Hague, M. Nijhoff, 1995).

Schiffler, Manuel, "Sustainable Development of Water Resources in Jordan: Ecological and Economic Aspects in a Long-Term Perspective" in J. A. Allan and Chibli Mallat (eds.), *Water In The Middle East: Legal, Political and Commercial Implications* (London, I. B. Tauris Publishers, 1995), p. 239.

Schoenborn, Brian J., "Public Participation in Trade Negotiations: Open Agreements, Openly Arrived At?" (1995) 4 Minn. J. Global Trade 103.

Schreuer, Christoph, "The Waning of the Sovereign State: Towards a New Paradigm for International Law?" (1993) 4 Eur. J. Int'l L. 447.

Schrijver, Nico, *Sovereignty over Natural Resources: Balancing Rights and Duties* (Cambridge, Cambridge University Press, 1997).

Schroeder, Christopher H., "Rational Choice Versus Republican Moment Explanations for Environmental Laws, 1969–73" (1998) 9 Duke Envtl. L. & Pol'y F. 29.

Schwabach, Aaron, "The Sandoz Spill: The Failure of International Law to Protect the Rhine from Pollution" (1989) 16 Ecology L.Q. 443.

Schwartz, Alan, "Relational Contracts in the Courts: An Analysis of Incomplete Agreements and Judicial Strategies" (1992) 21 J. Leg. Stud. 271.

Schwartz, Warren F. and Alan O. Sykes, "The Economics of the Most Favored Nation Clause" in Jagdeep S. Bhandari and Alan O. Sykes (eds.), *Economic Dimensions in International Law* (New York, Cambridge University Press, 1997), p. 43.

Schwebel, Stephen, "Second Report on the Law of Non-Navigational Uses of International Watercourses," reprinted in (1980) 2 *ILC Yearbook* (Part 2) 132.

Scott, Robert E., "Conflict and Cooperation in Long-Term Contracts" (1987) 75 Calif. L. Rev. 2005.

Sebenius, James K., "Negotiation Analysis" in Victor A. Kremenyuk (ed.), *International Negotiations: Analysis, Approaches, Issues* (San Francisco, Jossey-Bass Publishers, 1991), p. 203.

  Negotiating the Law of the Sea (Cambridge, MA, Harvard University Press, 1984).

Shelton, Dinah, "Human Rights, Environmental Rights, and the Right to Environment" (1991) 28 Stan. J. Int'l L. 103.

Shifrin, Robin, "Not by Risk Alone: Reforming EPA Research Priorities" (1992) 102 Yale L.J. 547.

Shugart, Matthew Soberg and John M. Carey, *Presidents and Assemblies* (Cambridge, Cambridge University Press, 1992).

Simms, Richard A.. "Equitable Apportionment – Priorities and New Uses" (1989) 29 Nat. Res. J. 549.

Sinclair, Ian, *The International Law Commission* (Cambridge, Grotius, 1988).

Singer, Michael, "Jurisdictional Immunity of International Organizations: Human Rights and Functional Necessity Concerns" (1996) 36 Va. J. Int'l L. 53.

Siy, Robert Y., Jr. (ed.), *Community Resource Management: Lessons from the Zanjera* (Quezon City, Philippines, University of the Philippines Press, 1982).

Skutel, H. J., "Turkey's Kurdish Problem" (1988) 17(1) *International Perspectives* 22.

Slaughter, Anne-Marie, "International Law in a World of Liberal States" (1995) 6 Eur. J. Int'l L. 503.

"A Typology of Transjudicial Communications" (1994) 29 U. Rich. L. Rev. 99.

"Governing the Global Economy through Government Networks" in Michael Byers (ed.), *The Role of Law in International Politics* (Oxford, Oxford University Press, 2000).

(Anne-Marie Burley), "Toward an Age of Liberal Nations" (1992) 33 Harv. Int'l L.J. 393.

(Anne-Marie Burley), "Law among Liberal States: Liberal Internationalism and the Act of State Doctrine" (1992) 92 Colum. L. Rev. 1907.

Slaughter, Anne-Marie, Andrew S. Tulumello, and Stepan Wood, "International Law and International Relations Theory: A New Generation of Interdisciplinary Scholarship" (1998) 92 AJIL 367.

Smith, Herbert Arthur, *The Economic Uses of International Rivers* (London, P. S. King & Son, Ltd., 1931).

Soloway, Julie A., "Environmental Trade Barriers under NAFTA: The MMT Fuel Additives Controversy" (1999) 8 Minn. J. Global Trade 55.

Stairs, Kevin and Peter Taylor, "Non-Governmental Organizations and the Legal Protection of the Oceans: A Case Study" in Andrew Hurrell and Benedict Kingsbury (eds.), *The International Politics of the Environment* (Oxford, Clarendon Press, 1992), p. 110.

Stevens, Georgina G., *Jordan River Partition* (Stanford, CA, Hoover Institution on War, Revolution, and Peace, Stanford University, 1965).

Stevens, Jane Ellen, "Science and Religion; Cultural Practices and Ecology" (1994) 44(2) *Bioscience* 60.

Stigler, George J., "The Theory of Economic Regulation" (1971) 2 Bell J. Econ & Mngm't Sci. 3.

Stone, Katherine Van Wezel, "Labor and the Global Economy: Four Approaches to Transnational Labor Regulation" (1995) 16 Mich. J. Int'l. L. 987.

Sturgess, Gary L., "Transborder Water Trading among the Australian States" in Terry L. Anderson and Peter J. Hill (eds.), *Water Marketing – The Next Generation* (Lanham, MD, Rowman & Littlefield, 1997), pp. 127–45.

Subedi, Surya P., "Hydro-Diplomacy in South Asia: The Conclusion of the Mahakali and Ganges River Treaties" (1999) 93 AJIL 953.

Sunstein, Cass R., *The Partial Constitution* (Cambridge, MA, Harvard University Press, 1993).

Szekely, Alberto, "'General Principles' and 'Planned Measures' Provisions in International Law Commission's Draft Articles on the Non-Navigational Uses of International Watercourses: A Mexican Point of View" (1992) 3 Colo. J. Int'l Envt'l L. & Pol'y 93.

Tajfel, Henry, "Experiments in Intergroup Discrimination" (1970) 223 *Scientific American* 96.

Tamrat, Imeru, *Constraints and Opportunities for Basin-wide Cooperation in the Nile – A Legal Perspective* in J. A. Allen and Chibli Mallat (eds.), *Water in the Middle East: Legal, Political, and Commercial Implications* (London, I. B. Tauris Publishers, 1995).

Tang, Shui Yan, *Institutions and Collective Action: Self-Governance in Irrigation* (San Francisco, ICS Press, 1992).

Tarlock, A. Dan, "Safeguarding International River Ecosystems in Times of Scarcity" (2000) 3 U. Denv. Water L. Rev. 231.

"Exclusive Sovereignty Versus Sustainable Development of a Shared Resource: The Dilemma of Latin American Rainforest Management" (1997) 32 Tex. Int'l L.J. 37.

"The Role of Non-Governmental Organizations in the Development of International Environmental Law" (1993) 68 Chi. Kent L. Rev. 61.

"The Law of Equitable Apportionment Revisited, Updated, and Restated" (1985) 56 U. Col. L. Rev. 381.

Taylor, Michael, *The Possibility of Cooperation* (Cambridge, Cambridge University Press, 1987).

Teclaff, Ludwick A., "Evolution of the River Basin Concept in National and International Law" (1996) 36 Nat. Res. J. 359.

"Fiat or Custom: The Checkered Development of International Water Law" (1991) 31 Nat. Res. J. 46.

Water Law in Historical Perspective (Buffalo, NY, W. S. Hein, 1985).

The River Basin in History and Law (The Hague, M. Nijhoff, 1967).

Templeman, Lord, "Treaty-Making and the British Constitution" (1991) 67 Chi. Kent L. Rev. 459.

Thesiger, Wilfred, *The Marsh Arabs* (London, Longmans, 1964).

Thomas, Gregory A., "Conserving Aquatic Biodiversity: A Critical Comparison of Legal Tools for Augmenting Streamflows in California" (1996) 15 Stan. Envtl. L.J. 3.

Thompson, Alexander, "Unilateral Enforcement in the Shadow of International Institutions: Canada, Spain, and the Northwest Atlantic Fisheries Organization" (paper prepared for presentation at the Annual Meeting of the American Political Science Association, September 1998).

Thompson, Barton H., Jr., "Water Markets and the Problem of Shifting Paradigms" in Terry L. Anderson and Peter J. Hill (eds.), *Water Marketing – The Next Generation* (Lanham, MD, Rowman & Littlefield, 1997), pp. 1, 22.

"Institutional Perspectives on Water Policy and Markets" (1993) 81 Calif. L. Rev. 671.

Topping, Audrey R., "Ecological Roulette: Damning the Yangtze" *Foreign Affairs* (Fall 1995).

Toye, Patricia (ed.), *Palestine Boundaries 1833–1947* (4 vols., Slough, Archive Editions, 1989), vol. IV, p. 602.

Trachtman, Joel P., "L'Etat c'est Nous: Sovereignty, Economic Integration and Subsidiarity" (1992) 33 Harv. Int'l L.J. 459.

Trebilcock, Michael J. and Robert Howse, *The Regulation of International Trade* (2nd edn, London, Routledge, 1999).

Trimble, Phillip R., "Globalization, International Institutions, and the Erosion of National Sovereignty and Democracy" (1997) 95 Mich. L. Rev. 1944.

"A Revisionist View of Customary International Law" (1986) 33 UCLA L. Rev. 665.

Tversky, Amos and Daniel Kahneman, "Judgment under Uncertainty: Heuristics and Biases" in Daniel Kahneman, Paul Slovic, and Amos Tversky (eds.), *Judgment under Uncertainty* (Cambridge, Cambridge University Press, 1982), p. 3.

Ullmann-Margalit, Edna, *The Emergence of Norms* (Oxford, Clarendon Press, 1977).

Underdal, Arild, "The Outcomes of Negotiation" in Victor A. Kremenyuk (ed.), *International Negotiations: Analysis, Approaches, Issues* (San Francisco, Jossey-Bass Publishers, 1991), p. 100.

Vagts, Detlev F., "Comment: The Exclusive Treaty Power Revisited" (1995) 89 AJIL 40.

Vicuña, Francisco Orrego, *The Exclusive Economic Zone* (Cambridge, Cambridge University Press, 1989).

Vierdag, E. W., "The Legal Nature of the Rights Granted by the International Covenant on Economic, Social and Cultural Rights" (1978) 9 Netherlands Yb. Int'l L. 69.

Vinogradov, Sergei V., "Observations on the ILC's Draft Rules: 'Management and Domestic Remedies'" (1992) 3 Colo. J. Int'l Envt'l L. & Pol'y 235.

Viscusi, W. Kip, *Rational Risk Policy: The 1996 Arne Ryde Memorial Lectures* (Oxford, Clarendon Press, 1998).

Fatal Tradeoffs: Public and Private Responsibilities for Risk (New York, Oxford University Press, 1992).

Wade, Robert, *Village Republics: Economic Conditions for Collective Action in South India* (Cambridge, Cambridge University Press, 1988).

"The Management of Common Property Resources: Collective Action as an Alternative to Privatisation or State Regulation" (1987) 11 Cambridge J. of Econ. 95.

Wagner, Wendy E., "The Science Charade in Toxic Risk Regulation" (1995) 95 Colum. L. Rev. 1613.

Walker, Vern R., "Keeping the WTO from Becoming the 'World Trans-science Organization': Scientific Uncertainty, Science Policy, and Factfinding in the Growth Hormone Dispute" (1998) 31 Cornell Int'l L.J. 251.

Waltz, Kenneth, *Man, The State and War: A Theoretical Analysis* (New York, Columbia University Press, 1959).

Ward, Hugh, "The Risks of a Reputation for Toughness: Strategy in Public Goods Provision Problems Modelled by Chicken Supergames" (1987) 17 Brit. J. Pol. Sci..

Waterbury, John, "Hydropolitics of the Nile Valley" in *Contemporary Issues in the Middle East* (Syracuse, NY, Syracuse University Press, 1979).

Weil, Prosper, *Perspectives du droit de la delimitation maritime* (Paris, Pedone, 1988).

Weisbrod, Burton A., "The Future of the Nonprofit Sector: Its Entwining with Private Enterprise and Government" (1997) 16 J. Pol'y Analysis & Mgmt. 541.

Weiss, Edith Brown, *In Fairness to Future Generations: International Law, Common Patrimony and Intergenerational Equity* (Tokyo, Japan, United Nations University, 1989).

White, T. Anderson and C. Ford Runge, "The Emergence and Evolution of Collective Action: Lessons from Watershed Management in Haiti" (1995) 23 *World Development* 1683.

Whiteman, Marjorie M., *Digest of International Law* (15 vols., Washington, DC, US Dept. of State, 1964), vol. III.

Whittington, Dale and Elizabeth McClelland, "Opportunities for Regional and International Cooperation in the Nile Basin" (1992) 17 *Water International* 144.

Wiener, Jonathan Baert and John D. Graham, "Resolving Risk Tradeoffs" in John D. Graham and Jonathan Baert Wiener (eds.), *Risk Versus Risk: Tradeoffs in Protecting Health and the Environment* (Cambridge, MA, Harvard University Press, 1995), p. 226.

Williamson, Oliver E., *The Economic Institutions of Capitalism: Firms, Markets, Relational Contracting* (New York, Free Press, 1985).

Wintrobe, Ronald, "Modern Bureaucratic Theory" in Dennis C. Mueller (ed.), *Perspectives on Public Choice* (New York, Cambridge University Press, 1995), p. 429.

Wirth, David A., "The Role of Science in the Uruguay Round and NAFTA Trade Disciplines" (1994) 27 Cornell Int'l L.J. 817.

Wittfogel, Karl A., *Oriental Despotism: A Comparative Study of Total Power* (New Haven CT, Yale University Press, 1957).

Wolfrom, Marc, *L'utilisation à des fins autres que la navigation des eaux des fleuves, lacs et canaux internationaux* (Paris, A. Pedone, 1964).

The Work of International Law Commission (5th edn, New York, United Nations, 1996).

World Commission on Environment and Development, *Our Common Future* ('The Brundtland Report') (Oxford, Oxford University Press, 1987).

World Bank, *World Development Report (Development and the Environment)* (New York, NY, Oxford University Press, 1992).

World Disasters Report (Published by the International Federation of Red Cross and Red Crescent Societies 1999).

Wouters, Patricia K., "Allocation of the Non-Navigational Uses of International Watercourses: Efforts at Codification and the Experience of Canada and the United States" (1992) 30 Can. Yb. Int'l L. 43.

Young, Oran R. "The Effectiveness of International Institutions: Hard Cases and Critical Variables" in James N. Rosenau and Ernst-Otto Czempiel (eds.), *Governance without Government: Order and Change in World Politics* (Cambridge, Cambridge University Press, 1992), p. 160.

International Cooperation (Ithaca, NY, Cornell University Press, 1989).

"The Rise and Fall of International Regimes" (1982) 36 Int'l Org. 277.

Bargaining: Formal Theories of Negotiation (Urbana, University of Illinois Press, 1975).

Index

Books in the series

Principles of the institutional law of international organisations
  C. F. Amerasinghe

Fragmentation and the international relations of micro-states
  Jorri Duursma

The Polar regions and the development of international law
  Donald R. Rothwell

Sovereignty over natural resources
  Nico Schrijver

Ethics and authority in international law
  Alfred P. Rubin

Religious liberty and international law in Europe
  Malcolm D. Evans

Unjust enrichment
  Hanoch Dagan

Trade and the environment
  Damien Geradin

The changing international law of high seas fisheries
  Francisco Orrego Vicuña

International organizations before national courts
  August Reinisch

The right to property in commonwealth constitutions
  Tom Allen

Trusts: A comparative study
  Maurizio Lupoi

On civil procedure
  J. A. Jolowicz

Good faith in European contract law
  Reinhard Zimmerman and Simon Whittaker

Money laundering
  Guy Stessens

International law in antiquity
  David J. Bederman

The enforceability of promises in European contract law
  James Gordley

International commercial arbitration and African states
  Amazu A. Asouzu